Evolution of Smart Sensing Ecosystems with Tamper Evident Security

Pawel Sniatala • S. S. Iyengar
Sanjeev Kaushik Ramani

Evolution of Smart Sensing Ecosystems with Tamper Evident Security

Pawel Sniatala
Department of Computing Science
Poznań University of Technology
Poznan, Poland

S. S. Iyengar
Knight Foundation School of Computing
and Information Sciences, CASE 351
Florida International University
Miami, FL, USA

Sanjeev Kaushik Ramani
Knight Foundation School of Computing
and Information Sciences, CASE 351
Florida International University
Miami, FL, USA

ISBN 978-3-030-77766-1   ISBN 978-3-030-77764-7 (eBook)
https://doi.org/10.1007/978-3-030-77764-7

© The Editor(s) (if applicable) and The Author(s), under exclusive license to Springer Nature Switzerland AG 2022

This work is subject to copyright. All rights are solely and exclusively licensed by the Publisher, whether the whole or part of the material is concerned, specifically the rights of translation, reprinting, reuse of illustrations, recitation, broadcasting, reproduction on microfilms or in any other physical way, and transmission or information storage and retrieval, electronic adaptation, computer software, or by similar or dissimilar methodology now known or hereafter developed.

The use of general descriptive names, registered names, trademarks, service marks, etc. in this publication does not imply, even in the absence of a specific statement, that such names are exempt from the relevant protective laws and regulations and therefore free for general use.

The publisher, the authors, and the editors are safe to assume that the advice and information in this book are believed to be true and accurate at the date of publication. Neither the publisher nor the authors or the editors give a warranty, expressed or implied, with respect to the material contained herein or for any errors or omissions that may have been made. The publisher remains neutral with regard to jurisdictional claims in published maps and institutional affiliations.

This Springer imprint is published by the registered company Springer Nature Switzerland AG
The registered company address is: Gewerbestrasse 11, 6330 Cham, Switzerland

Dedication to Prof. Asad Madni

Dr. Asad M Madni is a distinguished adjunct professor and distinguished scientist of electrical and computer engineering at the University of California, Los Angeles. Previously, he was president, COO, and CTO of BEI Technologies Inc., and chairman, CEO, and CTO of Systron Donner Corporation. He led the development and commercialization of intelligent sensors, systems, and instrumentation, for which he received worldwide acclaim, including the servo control system for Hubble Space Telescope's Star Selector which provided the it with unprecedented pointing accuracy and stability, resulting in truly remarkable images that have enhanced our understanding of the universe, and the revolutionary MEMS GyroChip® technology which is used worldwide for electronic stability control and rollover protection in passenger vehicles, thereby saving millions of lives every year. Dr. Iyengar is most grateful to him for his inspirational mentoring towards his professional career during the last 8 years, and his dedication and energy towards the *development of technologies for the future. Collaborations with him have yielded important high-impact results in several areas of sensing and computing. The authors are also very thankful for his support during this endeavor of compiling and presenting ideas of global importance in the area of evolution of smart sensing ecosystems and the need for tamper-evident security.*

Dedication to Prof. Daniel Berg
Dr. Daniel Berg, former co-provost at CMU and president of Rensselaer Polytechnic Institute and now affiliated with University of Miami.

I am very thankful for all the contributions he has made, and I benefitted having many inspiring and fruitful discussions and contributions to the technology of the future. It's very rare to find a distinguished person like Dr. Berg who has so much of energy and enthusiasm to mentor colleagues in this area of technology transition.

The authors are very thankful for his support during this endeavor of compiling and presenting ideas of global importance in the area of evolution of smart sensing ecosystems and the need for tamper evident security.

— S. S. Iyengar

Foreword

The mission of this book is to explain the evolution of techniques and strategies in securing information transfer and storage, thus facilitating a digital transition towards building tamper evident systems. The goal is also to aid business organizations that are dependent on the analysis of the large volumes of generated data in securing and addressing the associated growing threat of attackers relentlessly waging attacks and the challenges in protecting the confidentiality, integrity, and provenance of data.

The book provides a comprehensive insight into the secure communication techniques and tools that have evolved and the impact they have had in supporting and flourishing business through the cyber era. This book also includes chapters that discuss the most primitive encryption schemes to the most recent use of homomorphism in ensuring the privacy of data, thus leveraging greater use of new technologies like cloud computing and others.

The *Evolution of Tamper evident Secure Systems for the Cyber Era* is a book that provides a wealth of information for an audience that is involved in the development and management of information systems. This book gives industry persons, researchers, and students interested in securing their systems with the necessary information and tools that can be used to design a tamper evident and secure ecology in this highly vulnerable cyber era.

Vice Rector, Poznan University of Technology
Poznan, Poland

Pawel Sniatala

Distinguished University Professor,
Florida International University
Miami, FL, USA

S. S. Iyengar

Graduate Research Scholar,
Florida International University
Miami, FL, USA
April 2021

Sanjeev Kaushik Ramani

Preface

To understand this new tamper-evident data security protocol, let's first look at the traditional ways to data security and integrity. The typical way of safeguarding data is through the use of cryptographical methods which include encryption and decryption. Most people will have received an encrypted email at some point—an email that displays as a bunch of symbols, but if you are the intended recipient, a simple "key" or code unlocks the ciphertext (uninterpretable text) and you are able to read the message.

Cybersecurity trends and market studies have seen that modern businesses lose close to USD 100 billion every quarter in remediation of issues caused by cyberattacks. Data becoming highly available and pervasive has left the traditional security approaches requiring a paradigm shift to sustain. Offloading data security and maintenance to third parties is not a solution anymore. There are various points from where information can be tapped and hence identifying end-to-end security is an area requiring deep exploration and investigation. What kind of data is being sought after by the attackers is an unanswerable question because it is difficult to classify data into a binary group stating "useful" and "not-useful". Modern technology has grown to an extent where with very minimal context, information can be synthesized and even used to cause havoc. Deepfakes is a classic example where a novice attacker can follow instructions from the Internet to mask either audio or video and make the presence of others felt in a scene. The multiple attacks on cryptocurrencies have also brought into light the fragile and highly chaotic cyber world that we live in.

The existing and possible new gaps/holes in this field has to be fixed to keep a skilled cyberterrorist at bay because they can design a mechanism that can filter and break the security algorithms and reveal the information. Cyber attackers are getting smarter and more creative every day and we're always looking at ways to combat their efforts. The new technique we explore and propose in this book plays games with hackers and confuses them by using a variant of cryptography that combines multiple encryption models. On an abstract level, the tamper evident system randomly switches between multiple cryptographic algorithms, thus preventing hackers from identifying the algorithm used for encryption and thus break the ciphertext.

We have also identified the benefits of merging the homomorphic encryption technique and probabilistic encryption technique and compared it with the use of a single traditional approach. The real-world applications for the tamper evident system is manyfold, including storage and recovery from cloud operators, handling transmission of personally identifiable information (PII) in healthcare, tracking the dynamic changes in stocks, securing email chains, and protecting information in mobile devices and social media to name a few. The greatest requirement in the modern age is the need for empowering security that will provide tamper evident information exchange. This brings up a prominent need for techniques that integrate modern cryptosystem solutions to achieve enhanced information integrity and confidentiality with minimum impact to system performance. Cryptosystems have evolved leaps and bounds striving to achieve this. However, a paradigm shift is to be unearthed with the great pervasiveness of the attacker knowledge and attack surfaces. Generation and use of large volumes of data has made this a topic of great importance especially in this era. This book gives the reader a tour through the evolution of cryptosystems, their achievements, and the growth of more sophisticated systems that have affordable time and space complexities. Our mission is to explain the evolution of techniques and strategies in securing information transfer and storage, thus facilitating a digital transition towards the modern tamper evident systems. We hope this book can aid business organizations that are dependent on the analysis of the large volumes of generated data in securing and addressing the associated growing threat of attackers relentlessly waging attacks and the challenges in protecting the confidentiality, integrity, and provenance of data. This book provides a comprehensive insight into the secure communication techniques and tools that have evolved and the impact they have had in supporting and flourishing the business through the cyber era. This book also includes chapters that discuss the most primitive encryption schemes to the most recent use of homomorphism in ensuring the privacy of the data, thus leveraging greater use of new technologies like cloud computing and others. We, thus, hope that this book, titled *The Evolution of Tamper evident Secure Systems for the Cyber Era*, is the perfect ally for industry experts, researchers, students, and others who are interested in securing their information and systems from the prying eyes of attackers. This book provides a wealth of information for the audience with the necessary information and tools that can be used to design a tamper evident and secure ecology in this highly vulnerable cyber era.

Poznan, Poland	Pawel Sniatala
Miami, FL, USA	S. S. Iyengar
Miami, FL, USA April 2021	Sanjeev Kaushik Ramani

Acknowledgments

This work has evolved from the research conducted by the authors and other collaborators on understanding the security needs for the current information age where there is an explosion of information which if not secured and used optimally can lead to catastrophic outcomes. The golden era of information or the cyber age that we live in requires advancement in the areas of protocol design, technology transition to provide tamper evident solutions to all computing architectures and consumers of technology. This book reports an exploratory attempt to highlight the evolution of smart systems that we use on a day-to-day basis and the need for securing such devices. The book also explores the existing techniques and analyzes the shortcomings and critical gaps and proposes a variant that combines the benefits of the existing algorithms on different planes in identifying a tamper evident approach to securing information and also identifies future directions that will occupy this space in times to come.

Many people made this book possible and contributed to it, whom the authors are very grateful to.

This book benefited from the work of and input of a range of people, whom the authors gratefully acknowledge here. From the funding point of view, the authors would like to thank the US National Science Foundation for funding for several years, the US Army Research Office, and early funding from the Office of Naval research during the 1990s to early 2000s.

The authors would like to acknowledge the support of Professor Teofil JESIONOWSKI (Rector, Poznan University of Technology) for his strong support of the collaboration between Florida International Unviersity and Poznan University of Technology. The authors would also like to thank Dr. Latesh, Dr. Mario Mastriani, Col. Jerry Miller, and other friends, colleagues, and collaborators who have directly or indirectly contributed towards the completion of this book. The authors would like to thank their families for the support extended during the compilation of this book.

Contents

Part I Growth of Sensory Devices and Their Proliferation into Daily Life

1 Introduction 3
    1.1   Sensing and Classification 4
    1.2   Growth of Wireless Sensor Networks 5
    1.3   Smart Sensing Devices: Why Do We Need Them? 6
    1.4   Automation of Processes: Aided by Sensors 7
    1.5   Final Remarks 8
    Reference 8

2 Smart Objects in Cyber-Physical Systems 9
    2.1   Cyber Physical Systems (CPS) 10
    2.2   The Internet-of-Things (IoT) Paradigm 11
    2.3   The Boom of Wearables, Handhelds, and Smart Devices 11
    2.4   Services Regalia: Touching the Lives of Millions 13
    2.5   Volumes of Data Generation: Boon or Bane? 15
    2.6   Final Remarks 15
    References 16

3 IoT Security 17
    3.1   Why Do We Need to Secure Smart Devices? 18
            3.1.1   Device Bootstrapping 18
            3.1.2   Authentication 19
            3.1.3   Access Control 20
    3.2   Challenges to Securing Smart Devices 21
    3.3   Can IoT Devices Be Used to Wage Attacks: The Rise
           of Botnets 22
    3.4   Techniques to Protect Devices 23
    3.5   Concluding Remarks 24
    References 24

Part II Securing the Internet-of-Things Enabled Smart Environments

4 Understanding the Smart Systems: IoT Protocol Stack 27
- 4.1 Introduction 27
- 4.2 Current Internet Protocols: Can They Survive the Meteoric Rise of IoT Device Adoption? 29
- 4.3 Available Standards 30
- 4.4 The Six Layers in IoT Security 32
  - 4.4.1 Securing the Network of Operation 33
  - 4.4.2 Authentication Requirements of Devices 33
  - 4.4.3 Encrypting Data to Prevent Leakage of Information in Clear Text 33
  - 4.4.4 Storage Solutions 34
  - 4.4.5 Device Lifecycle Management 34
  - 4.4.6 Interface or API Protection 35
- 4.5 Concluding Remarks 35
- References 35

5 Onboarding New IoT Devices 37
- 5.1 Talking to the Correct Device/Network?: The Need for Device Bootstrapping 37
- 5.2 Taxonomy of Out-of-Band (OOB) Approaches 39
  - 5.2.1 Bluetooth Low Energy (BLE) 39
  - 5.2.2 Haptics/Touch 40
  - 5.2.3 Magnetic Field Technique 41
  - 5.2.4 Visual Techniques 41
  - 5.2.5 Audio Techniques 41
  - 5.2.6 Vibration Techniques 42
- 5.3 Existing Bootstrapping Techniques 42
- 5.4 Concluding Remarks 42
- References 43

Part III Cryptosystems: Foundations and the Current State-of-the-Art

6 Introduction 47
- 6.1 Motivation 47
- 6.2 Cryptographic Solutions to Secure Information 48
- 6.3 History of Cryptography 48
- 6.4 Current State-of-the-Art 49
- 6.5 Issues with Existing Techniques 49
- 6.6 Do We Have a Solution? 52
- 6.7 Final Remarks 53
- References 53

7 Symmetric Key Cryptography ... 55
- 7.1 Applications and Advantages ... 55
- 7.2 Stream Ciphers ... 56
- 7.3 Block Ciphers ... 56
  - 7.3.1 Initialization Vectors ... 57
- 7.4 Major Disadvantages of Symmetric Key Cryptosystems ... 58
- 7.5 Cryptographic Attacks on Symmetric Key Cryptosystems ... 58
  - 7.5.1 Key Search (Brute Force) Attacks ... 59
  - 7.5.2 Cryptanalysis ... 59
  - 7.5.3 Systems Based Attack ... 60
- 7.6 Final Remarks ... 61
- References ... 61

8 Asymmetric Key Cryptography ... 63
- 8.1 What Is Public Key Cryptosystem? ... 63
- 8.2 Advantages of Asymmetric Key Cryptosystems ... 64
- 8.3 Drawbacks of PKI ... 64
- 8.4 Concluding Remarks ... 65

Part IV Modern Encryption Schemes

9 Homomorphic Encryption ... 69
- 9.1 Applications of Homomorphic Encryption ... 70
  - 9.1.1 Data Security in the Cloud ... 70
  - 9.1.2 Support to Data Analysis ... 72
  - 9.1.3 Enhancing Secure Ballot and Electoral Systems ... 73
- 9.2 Classification of Homomorphic Encryption: Examples ... 74
  - 9.2.1 Partially Homomorphic Encryption (PHE) ... 75
  - 9.2.2 Somewhat Homomorphic Encryption (SHE) ... 75
  - 9.2.3 Fully Homomorphic Encryption (FHE) ... 76
- 9.3 Problems of Importance in the Public Key Infrastructure ... 76
- 9.4 Concluding Remarks ... 76
- References ... 76

10 Popular Homomorphic Encryption Schemes ... 77
- 10.1 Goldwasser–Micali Method (GM Method) ... 77
  - 10.1.1 Encryption ... 78
  - 10.1.2 Decryption ... 79
- 10.2 Paillier Encryption Scheme ... 80
  - 10.2.1 Benefits of the Additive Homomorphic Encryption ... 80
- 10.3 ElGamal Encryption Scheme ... 81
- 10.4 RSA Cryptosystem ... 82
- 10.5 Concluding Remarks ... 83
- References ... 84

11 Industrial Involvement in the Use of Homomorphic Encryption 85
- 11.1 IBM's HElib 86
- 11.2 Microsoft SEAL 86
- 11.3 Other Homomorphic Encryption Library 87
- 11.4 Final Remarks 88
- References 88

Part V Creating a Tamper Evident System for the Cyber Era

12 Introduction 91
- 12.1 Limitations of Existing Public Key Encryption Techniques 91
  - 12.1.1 Major Limitations 92
  - 12.1.2 Can a Shift from Deterministic to Probabilistic Approaches Provide a Better Outcome? 92
- 12.2 Motivation: Our Major Proposition 93
- 12.3 Tamper Evident Solutions 93
  - 12.3.1 Working of the Proposed Scheme 94
  - 12.3.2 Advantages of TEDSP 94
- 12.4 Impact on Information Storage and Retrieval Applications 94
- 12.5 Final Remarks 95

Part VI Feasibility and Performance Analysis

13 Implementation Details 99
- 13.1 Details of the Code Implemented 99
- 13.2 Comparison of Performance 101
- 13.3 Evaluation 103
- 13.4 Final Remarks 103
- Reference 104

Part VII Future Directions

14 Quantum Cryptography 107
- 14.1 Prolegomenous on Quantum Key Distribution (QKD) 108
- 14.2 A Primer on Quantum Information Processing 109
- 14.3 QKD Based on Polarized Single Photon 112
- 14.4 QKD Based on Entangled Photon Pairs 114
- 14.5 Final Remarks 115
- References 116

15 Quantum Tools 119
- 15.1 Quantum Entanglement 119
- 15.2 Quantum Teleportation 122
- 15.3 Modern Version of QKD Based on Entangled Photon Pairs 126
- 15.4 Virtual Entanglement Procedure 127
- 15.5 Quantum Internet 131

		15.5.1	Entanglement Swapping for Quantum Repeaters	131
		15.5.2	Quantum Repeaters for Quantum Internet	133
		15.5.3	Quantum Memories for Buffering	140
	15.6	Final Remarks		140
	References			141
16	**Applications**			**143**
	16.1	Quantum Internet of Things (QIoT)		143
	16.2	Quantum Blockchain		144
	16.3	Quantum Money and Quantum Cheque		145
	16.4	Quantum Security of Confidential Documents		146
	16.5	Quantum Radar		147
	16.6	Quantum Medical Imagery		148
	16.7	Final Remarks		149
Index				**151**

Acronyms

The list of abbreviations, symbols that are commonly used in this book are as listed below

AES	Advanced Encryption Standard
BLE	Bluetooth Low Energy
CA	Certification Authority
CBC	Cipher Block Chaining
CFB	Cipher Feedback
CPS	Cyber Physical Systems
CTR	Counter
DDoS	Distributed Denial of Service
DES	Data Encryption Standard
DoS	Denial of Service
FHE	Fully Homomorphic Encryption
GCM	Galois/Counter Mode
IoT	Internet of Things
IV	Initialization Vector
LIDAR	Light Detection and Ranging
OOB	Out of Band
OTP	One-Time Password
PHE	Partially Homomorphic Encryption
PKI	Public Key Infrastructure
RADAR	Radio Detection and Ranging
SHE	Somewhat Homomorphic Encryption
SMS	Simple Message Service
SONAR	Sound Navigation and Ranging
TLS	Transport Layer Security
Wi-Fi	Wireless Fidelity
WoT	Web of Trust

Part I
Growth of Sensory Devices and Their Proliferation into Daily Life

Chapter 1
Introduction

The rapid growth in technology and the interconnectivity we have achieved among devices is largely due to the availability of small, highly inexpensive sensory devices. These sensors have helped us realize and fuel the Internet-of-Things (IoT) paradigm and made it available at affordable prices to the common man. Using these devices, one can easily revolutionize their daily life and make their surroundings and environment smart. These devices have also been pivotal in transforming human life and bridging the gap between human and devices and even among heterogeneous devices. The great advancement in material science and engineering which gave the sensors we see today has given a new lease of life to the way things are perceived and this highly connected global ecosystem where every human and thing communicates and function seamlessly. Figure 1.1 depicts how common day-to-day devices can be interconnected to provide services and enhance the livelihood.

The first evidence of sensing is traced back to the 1800s when variable resistors were used to identify temperature fluctuations. The discovery of semiconductors and the vast applications that semiconductor physics unearthed has seen sensors to proliferate to an extent such that they are either embedded into the human body directly or even worn for continuous tracking and ignite vital human parameters and enable wearable computing. Enhanced manufacturing techniques have ensured that we can achieve all this at very meager costs.

The increasing computational capabilities of sensors and the demands of varied applications to use them have led to the growth of smart objects. These smart objects form the basis that helps in designing and building of the Cyber-Physical Systems (CPSs) and Internet of Things (IoT). In this chapter of the book, we shall give an exhaustive insight into the way sensors evolved and the advancements that led to the development of smart objects. We shall also discuss the growth of wireless sensor networks (WSNs) and further extend our discussions toward the growth of CPS and IoT. We shall also introduce the security aspects around these smart systems and possible countermeasures.

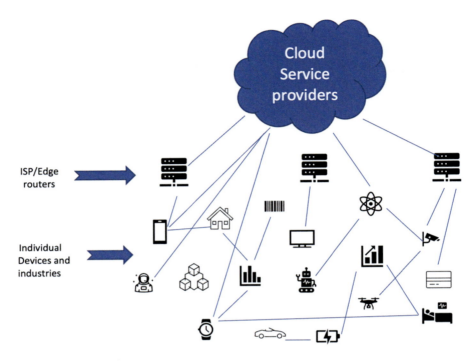

Fig. 1.1 A connected ecosystem

1.1 Sensing and Classification

The word sensor is commonly associated with any device that is capable of collecting, processing, and (or) transmitting information collected physically. The material used to build the sensor decides the quantity and aspect it measures. Sensing is the conversion of energy from one form to another (more often the output is a voltage value that can be measured and quantified). In most cases, sensors are not individually used but are packaged as a unit to give the user control over what is being sensed and how it can be processed.

A common method for classifying sensors is based on the forms of energy they work with or transform the sensed values into. Another classification technique involves the place of deployment of such sensors and the environment in which they operate. Multiple sensors are usually merged together to form a collaborative sensing unit/system and deployed based on the needs of the application the device is designed to do. A system that is well known and used in our day-to-day lives is the common smartphone which is equipped with multiple sensors (Fig. 1.2) to perform various operations. A subset of sensors that a commodity smartphone possesses includes accelerometers, gyroscopes, magnetometers, cameras, ambient light sensors, etc. Apart from these, a modern smart-home is equipped with

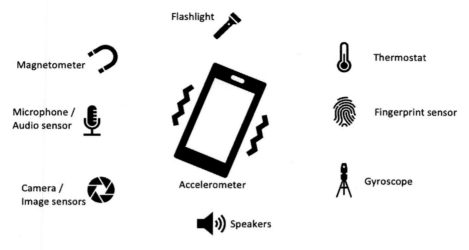

Fig. 1.2 Common sensors in a smartphone

biological sensors, humidity sensors, and chemical sensors that sense particulate matter in air, pH sensors for the fluids being consumed, surveillance devices including cameras, motion detectors, proximity sensors, and many more.

1.2 Growth of Wireless Sensor Networks

The large-scale use of sensors and sensory devices was fuelled by and largely supported by the possibility of wirelessly communicating among many of these devices. Wireless support made these devices highly portable and thus deployable and controllable even in remote areas. The subsequent growth of the wireless sensor network (WSN) technology brought about a radical change and opened up newer avenues and applications for these sensing devices. Through WSN, these devices could connect and share their sensed data and thus reduced the load on processing and storage at the sensing site. Application of newer techniques like crowdsourcing and other data cleaning techniques has provided the necessary push for these devices and networks.

A formal definition of wireless sensor networks identifies WSNs as a network of sensors that are distributed spatially and strategically with the intention of performing certain applications in the deployed environment [1]. The system is organized with a gateway/cluster head/aggregation node that connects the network back to the wired world and other distributed networks. A successfully deployed and coordinated WSN provides greater capabilities in terms of sensing performance, ease of maintenance, possible self-organization, reduction in the individual sensor sizes, highly granular sensing, collaborative benefits, etc. to name a few.

Every technology has a flip-side with drawbacks. Researchers over the years have identified and tried to solve the issues that WSNs possess. Security of the data being transmitted, trustworthiness of the information and the transmitting nodes, access control rights of the nodes and the cluster head, information fusion, and consensus techniques are some of the topics that we discuss in this book. There are other aspects as well that are being actively looked upon by researchers.

1.3 Smart Sensing Devices: Why Do We Need Them?

Now that we have identified and defined what sensors are, it is imperative to understand their need and use. The industrial revolution brought about many major changes and one important discovery is that of semiconductors and semiconductor physics. Soon, there came up a field of study related to semiconductor chip manufacturing and integration of many transistors to form a working system capable of performing computations. Advancements in this field led to the creation of systems on a single chip and providing multiple capabilities in devices that could measure in nanometers (nm).

A smart sensor is thus a by-product of such advancements and being capable of capturing environmental changes and processing them with very minimum errors. Such smart sensors, which could have a complex anatomy compared to a traditional sensor, are deployed in abundance in mission critical devices in warzones, advanced aviation techniques, efficient power distribution using smart power grids, patrolling and surveillance, space explorations, etc. The possibility of interfacing these devices through the Internet with unique identifiers (UIDs) and thus be a part of the global picture and making Internet of Things (IoT) a reality. The use of smart sensors instead of the traditional sensors has led to the growth of wireless sensor and actuation networks (WSANs).

Smart sensors in comparison with traditional sensors are advantageous by being more robust, providing easier maintenance, better calibration, and greater availability along with enhanced precision. Such devices gradually started being integrated into smart objects, which predominantly works with sensed information. It is thus very critical that we ensure the security of these devices and the data that is being collected, processed, and transmitted. Modern hardware-based security attacks have made it a topic of greater criticality to ensure that the devices are secured both physically and through software patches. The highly distributed nature in which these devices are placed and work makes it a highly sort after target for security attacks as it could give the attacker access to the overall networks and other peer networks and even lead to these compromised devices acting as bots in a larger cyber-attack.

1.4 Automation of Processes: Aided by Sensors

Through robotics, generation of feature-rich datasets for analysis, controlling electricity and water consumption, etc., sensors play an important role in automating everyday tasks. Sensors and smart devices are also being employed to perform critical tasks where humans cannot be present due to uninhabitable conditions. The outer space explorers and rovers, cooling systems in nuclear reactors, pacemakers that keep the patient's heart functioning, robotic low-invasive surgery, etc. are some of the many examples. Specialized sensors are being deployed in health monitoring and automated tracking of vital parameters, thus helping doctors connect through telemedicine and provide effective cures. Nanorobots and other sensory vectors are also used in effective administration of medicines.

State-of-the-art industrial automation lines use sensors all through their automated manufacturing chains starting from modeling, to assembling, to highly complex and critical tasks. The entire basis of autonomously driving vehicles is dependent on the environment sensing capabilities of the embedded sensors failure of which can have catastrophic outcomes. It is astonishing to know that the fault detection mechanisms in such systems also employ sensors and information fusion. The proliferation of IoT and CPS has enabled the use of sensors in appliances that do mundane activities and make the world a better place to live in. Studies have revealed that the use of appropriate sensors at appropriate locations can help reduce energy consumption by almost 40%, thus helping in reducing the carbon footprint and prevent global warming.

The intelligence that these sensors and networks of sensors possess is astounding. It is overwhelming to see the extent to which humans are dependent on them for their everyday tasks. This over-dependence could have adverse effects if control is lost to entities with malicious intentions. Cyber-attackers can utilize these sensors to wage many passive and active attacks, evidences of which have been seen in the news lately. This book highlights some of these issues and also provides a tamper evident model that can be employed to protect information from such devices even in this highly vulnerable cyber era.

Figure 1.3 depicts the possible automated services we can leverage using a sensor-infused smart-home. From the point of view of an attacker, these devices and the automated services are possible easy targets to control the house and thus perform malicious activities. As an example, the adversary can try to alter the temperature and radiator to such an extent that the smoke detector and alarm start buzzing and water starts gushing out of the sprinklers to douse a fire that does not actually exist. This scenario can get complicated if the window control and door control are set to open at the time of a fire alarm to allow fumes to escape the house, which will now provide the adversary physical access into the house to induce greater damage. Thus, it is important to ensure that effective measures are taken to monitor and control the information generation, transfer, actuation, and other storage of information. The use of cloud as a possible storage option may have other risks, which will be discussed in upcoming chapters of this book.

8   1   Introduction

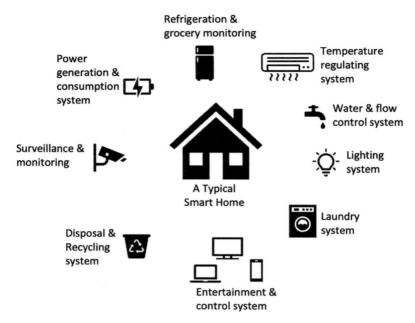

Fig. 1.3 A subset of sensors in a smart-home

1.5 Final Remarks

Sensors and actuators are slowly becoming important and irreplaceable elements of our daily life. Smartphone, which is a complex system with multiple seniors, is a common feature nowadays and is capable of generating large volumes of information that can be sensitive and private as well. It is thus highly important to be able to control these devices and the data they generate, use, transmit, and store in order to reduce the risk of being attacked by cyber-criminals. With IoT adopting these sensors and increasing our dependence on these smart sensors and actuators, a radical change to a tamper evident security system is inevitable.

Reference

1. Ramani, Sanjeev Kaushik, and S. S. Iyengar. "Evolution of sensors leading to smart objects and security issues in iot." In International Symposium on Sensor Networks, Systems and Security, pp. 125–136. Springer, Cham, 2017.

Chapter 2
Smart Objects in Cyber-Physical Systems

Smart sensor discovery and advancement in the manufacturing techniques led to the design and development of novel devices and techniques that could successfully bridge the gap between humans and machines. The devices with added intelligence are nowadays made part of complex systems that are often called as cyber-physical systems (CPSs) [1]. The smart objects that define these CPSs have redefined and revolutionized our communication to such an extent that they have sprung new life to our ways of computing and information sharing with greater emphasis on being content-driven and context-aware. Millions of these smart objects loosely work as a cluster of distributed networks that support IoT and enable the use of modern technologies like blockchains and other distributed ledgers.

The idea of a smart environment with smart objects was conceived by Mark D. Weiser, who is also the father of "ubiquitous computing." He envisioned all devices to have computational capabilities and be able to act according to the environment in which they are being operated by either sensing their surroundings or being able to collaboratively work with other devices in retrieving critical information for decision-making [2]. Smart objects since then have evolved and now are either incorporated with artificial intelligence and learning capabilities or even used as parts of complex systems to harness their characteristics which include context awareness, abstraction capabilities, communication interfaces and capabilities, proactive and reactive behavior to changes in the surroundings, etc. [3].

With the world having seen a plethora of these smart devices, they are coarsely categorized into the following:

- activity sensing—used for recording any changes to the environment;
- policy aware and bound—monitoring the surroundings for compliance with set policies and rules; and
- process aware—work toward a set goal by following step-by-step instructions.

The smart objects are also equipped with intelligence to be able to decide and send notifications to a controller/cluster head of any ambiguities in operation and even resolve issues.

An important aspect of the successful operation of these systems, objects, and devices is the information they can sense, collect, aggregate, retrieve, or communicate, and it is necessary for this information to be secured and follow adequate access control policies to ensure that it reaches the appropriate entities. This book discusses existing and possible attack vectors and surfaces and countermeasures to build a tamper evident system for this tangible information.

2.1 Cyber Physical Systems (CPS)

Cyber-physical systems is the generic term that is used to describe a system that involves feedback from either humans, environment, or the political state of the system which aids in networked or distributed, adaptive or predictive, intelligent and real-time systems that are resilient to cyber-attacks and with applications in communication, energy generation and distribution, health-care, automation, transportation, etc. supported by highly scalable and complex design and management. Active research is underway in this area with the US National Science Foundation (NSF) attributing CPS as a system that is engineered and works on seamless interaction and integration among computational algorithms and the physical systems [4].

CPS as touted by researchers is expected to revolutionize human interaction with man-made devices. Once extensively deployed, it will bring about the 4th Industrial Revolution. It will also have a great impact on the way in which information is retrieved, managed, and used for actuation. The advancement in the development of smart objects and AI fuels the CPS development and deployment into applications like smart-grids, autonomous systems, aviation, health-care, information-security, storage systems, communication methods, etc. [5].

The advent of CPS would add in more intelligence to the existing embedded systems, enhance their adaptability, efficiency, robustness, and safety, and make them highly scalable [6]. The challenges that we would have to overcome for the CPS to be extensively deployed include the variations in the design patterns followed by the various engineering disciplines, security of such devices, and the communication medium used by them. The ways to handle and the processing of highly voluminous big data that is generated by these systems and their security are also some of the challenges to be addressed. The popularity of smartphones has also encouraged the inroads that we have made in this field.

2.2 The Internet-of-Things (IoT) Paradigm

The current Internet architecture that has revolutionized the way information is exchanged is about five decades old since it was conceived. There have been many important advancements be it the World Wide Web (WWW), the rise and growth of cellular Internet from GPRS to edge to the current LTE (we are talking about 6G at the time of writing this book). However, the massive use of Internet was envisioned when things were given the intelligence to be able to connect to its peers and communicate over a network. The Internet-of-Things (IoT) paradigm as it is called today opened up a plethora of applications and opportunities for the devices.

IoT also brought about the first need for context-aware and content-aware computing. The computing world now considered the interconnectivity between highly heterogeneous devices that exchange real-time information with the need for real-time (on the fly) processing and rendering in some cases. Overall, there is the generation of a large volume of information through the many sensors that are a part of the IoT infrastructure, and it is important for us to create a way to secure it. In IoT, millions of smart devices are connected through the Internet. It is based on the vision of connecting and having access to all real-life physical devices including vehicles, buildings, utility devices, household items, control systems, and other things in real time over the Internet. All the above promises are made based on the rapid advancements made in sensor networks and radio frequency identification (RFID) technology making it an integral part of human life today [7].

The term IoT was conceived and coined in 1999 by the British visionary Kevin Ashton [8]. Since then, technology and market trends related to IoT have skyrocketed, and by 2025 according to a report from Mckincey Global Institute, the technology will be worth more than $6.2 trillion. Other studies have reported the use of about 50 billion IoT devices across the world by the end of 2020. However, technological challenges in terms of managing heterogeneous devices, interconnecting them, interoperability issues, and regulatory compliance have slowed the growth considerably. Among all the other issues, two main challenges that have to be addressed before we can orchestrate a full-fledged smart environment are security (physical and information) and the privacy of the generated data.

2.3 The Boom of Wearables, Handhelds, and Smart Devices

Wearable computing is a technology that is born out of the smart sensing and interconnecting (over the Internet) capabilities provided by IoT. This has transformed into a highly pervasive technology up to the extent that people rely on these devices and manage their day-to-day activities. Figure 2.1 shows a subset of wearable devices we use in our daily lives and the extent to which we are dependent on them. Another major revolution is the capabilities of the modern smartphones. Today's smartphones come equipped with many sensors and sensing units, which enhances

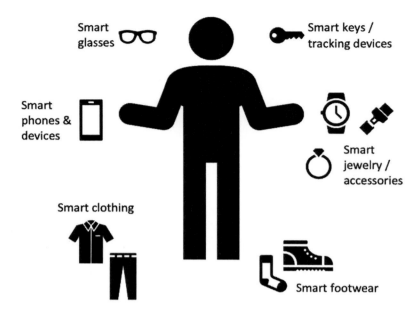

Fig. 2.1 Example of how pervasive wearable devices have become

the sensing capabilities. Advanced memory and processor design have given it an added boost to process and store large volumes of data with ease.

This unbridled technological advancement has also brought to life devices and objects that were once out of reach of human imagination. Applications of these devices have been identified for both the industry and personal use [9]. The year 2011 saw the discovery of a totally new dimension to the use of wearable devices and sensors with the launch of the Google glasses. This introduced the concept of virtual reality (VR) and augmented reality (AR) to the users in an immersive environment conducive for entertainment, gaming, etc. Since then, there have been many more application surfaces where VR/AR has been deployed with these technologies having the potential to simulate even the most complex war like scenarios to train military squadrons. The current versions of smartphones have evolved to be capable of rendering VR videos that the users can watch using specialized receivers/devices.

The Google glasses have evolved over time, and in 2015, a newer version of it was launched incorporating the features in a smartphone with the capability of Internet browsing, capturing photos and videos, navigation using maps, calendars, etc., all of which can be controlled easily using voice-based commands. However, these glasses soon came under the scanner of "Privacy" as everything in the vision of the user could be captured violating the privacy of others in the surroundings. Also, the hefty price tag and the management and processing of the tonnes of data collected were tedious and expensive leading to it being discontinued. However, it brought to light the endless possibilities of sensor technology and the immediate need for solutions that can ascertain security and privacy of information.

Another device that is rich with sensors and with many applications dependent on the sensing capabilities is the smartwatches. In 2015, Apple came up with their version of a smartwatch that can connect users to the world through an associated Apple iPhone. There are many other brands that have come up with smartwatches since and have been successful in gaining the attention of customers who can use these devices for viewing notifications, making and answering calls, reading and replying to texts, control other devices, access and perform operations on the applications in the connected phone, work as a fitness equipment, etc. There are other variants of wearable devices and can be broadly categorized based on their use as follows:

Devices for Monitoring Health
Among the estimated $3 trillion IoT industries by 2026, a major chunk is due to its use in healthcare devices. The devices used in this category range from personal trackers, constant vital monitoring, monitoring activities of a toddler to sustaining life when embedded as a pacemaker. With medical data being highly sensitive and private, it is of utmost importance to be able to prevent any kind of breach and compromise of such devices. A tamper evident system is thus required to secure these devices and the data they produce and consume.

Wearable Tags for Authorization
The industry is moving ahead of passwords and pins to grant access to employees and visitors into an infrastructure. IoT devices have given the capability of creating unique code similar to the user's biometrics to be used with a wearable device that can authenticate and thus provide access to the user. However, this technology is still in its nascency and needs a great amount of research and innovation to maintain the integrity and confidentiality of the information and also to thwart leaking and corruption of private information and the increasing threat of cyber-terrorism.

Corroborating institutional concern is a 2019 Worldwide Threat Assessment by the U.S. Intelligence Community warning of "financially motivated cyber-criminals" increasingly seeking to target critical infrastructure within the health care, financial, government, and emergency service sectors. Expect workers, travelers, and key personnel mandated to adopt wearable IoT technologies as the next frontier as society negotiates ever unpredictable outcomes and opportunities.

2.4 Services Regalia: Touching the Lives of Millions

The growth of sensing capabilities aided by the highly efficient sensors has led to the large-scale adoption of them in building IoT solutions and services. Billions of devices have been designed to be used in making this world a connected ecosystem with seamless interconnection among devices in geographically different locations supported by the Internet. Each of these devices is given enough intelligence for their application purpose and mimic a miniature computer and can ubiquitously work in the wireless environment. All of us are adept with the terms like smart-homes, smart-grids, etc. and have also seen autonomous driving vehicles that are loaded with millions of these small sensory units. It is fascinating that simple temperature and pressure sensors we see in our modern homes are the ones that handle the complex and sophisticated health of the aircraft (telemetry data) or even a space shuttle while it is flying thousands of miles into the atmosphere and beyond.

It is thus very evident that the sensors have a great impact on our lives. Every aspect of our lives is being fulfilled by these small technological workhorses. As an average consumer, we invariably resort to using the sensors in enabling our activities from sophisticated health care monitoring using wearable devices (watch, fitness tracker, etc. discussed above) to even systems that can help us track our missing phones, keys, or other devices, turn on/off devices in the kitchen, living spaces, or even a car to monitoring the growth of plants in the backyard or even farmlands, etc., the application surface is enormous with the limit being the extent to which our imagination can reach.

Sensors have enables us to enter this new digital age where the information collected be it service-specific data or even meta-data has helped us identify patterns/contexts that we were totally unaware of earlier. These patterns help us in constantly evolving and pushing the bar set for Quality of Service (QoS) and Quality of Experience (QoE) as a company/service provider. Newer techniques have come up with edge-, fog-, and dew-computing and the well-known cloud-based services that are capable of rapidly synthesizing the sensor data on the fly and provide a contextual output for the application domain. With the data collected being semantically very rich, it opens up doors for better and highly affordable services using the interconnection of devices. With the human imagination being the limit, there are new fictional TV shows that discuss the presence of a digital after-life where the human memories from when in earth can be uploaded so as to provide you with a living experience even after death in an alternate universe.

With over a trillion devices to be added to the connected devices kingdom across the globe, the generation of data has seen a boom over the past couple of years with more than 90% of the current information base being generated in the past decade. As an example, on an average, all the sensor-equipped devices in the world today are generating information at a rate such that a day's information is equivalent to the amount of data that was generated from when civilization started to the beginning of this millennium. Along with the favorable aspects/use of this data, it is very important for the information to be safeguarded appropriately as it can easily deem to be highly destructive if it fell into the wrong hands and were to be used maliciously.

2.5 Volumes of Data Generation: Boon or Bane?

Smart devices embedded with sensors are capable of generating enormous volumes of information. As much as we would like to control the activity of the microwave oven in the house using our smartphone or even a smartwatch, we should be aware that losing this control to a highly skilled attacker can lead to devastating and catastrophic outcomes as the attacker will be capable of even destroying the device and cause damage to our house and property. The sharing of information with service providers for state-of-the-art services and the recent breach of these information storage nodes exposing our personal and "private" information provide attackers to get a sense of our lives and even invade into our personal lives to be able to control it. The recent data breaches we have seen in this and the past decade have seen enormous loss monetarily and loss of reputation to large companies, and even the end-users (consumers) [10] identify a small subset of the major breaches that we have encountered since the dawn of the twenty-first century.

Even though it is highly lucrative to provide the sensory information to service providers in exchange of their services, it is always a good practice as a consumer to be aware of the consequences of sharing the information and the possible outcomes if the information was to be compromised at any point in time. The fine line between public and private information is to be drawn by all of us in deciding which information can be shared or is worth sharing and which is not. Also, with the lack of standards in the IoT realm for securing the information, it has exposed multiple loopholes that an attacker can take to wage attacks and compromise networks and systems. In conclusion, the opportunities that the enormous volumes of sensory information opens us to is very fascinating; however, it comes with the caveat that misuse/abuse of this information for malicious activities can have irreversible and highly catastrophic impacts. It is thus very important to understand the information being collected and shared and the security protocols that are in place to safeguard this information. The next chapter discusses the state-of-the-art in IoT security and the need for a paradigm shift in both the underlying Internet and the protocols used to secure the communication.

2.6 Final Remarks

The growth of smart devices and their adoption in our everyday life has seen a remarkable increase in the past decade. This has been because of the many features and value additions they give us and the inexpensive price points they come with which has been possible with the large-scale manufacturing processes. These sensors and devices are here to stay but bring with them a large number of potential problems and challenges that will have to be addressed and remediated before we can truly experience the connected world envisioned by IoT technology where every human and thing can communicate seamlessly without any hassles and securely exchange information. The following chapters describe IoT devices in more detail with specific focus on the state-of-the-art security of such devices, the challenges, existing and potential need for newer solutions.

References

1. Fortino, Giancarlo, Trunfio, Paolo (Eds.) (2014), Internet of Things Based on Smart Objects, Technology, Middleware and Applications, ISBN 978-3-319-00491-4.
2. Mark Weiser (1991), The Computer for the 21st Century, Scientific American (September 1991), 265, 94–104 https://doi.org/10.1038/scientificameri-can0991-94
3. G. Kortuem, F. Kawsar, V. Sundramoorthy and D. Fitton (2010) Smart objects as building blocks for the Internet of Things IEEE Internet Computing, vol. 14, no. 1, pp. 44–51, Jan.-Feb. 2010. https://doi.org/10.1109/MIC.2009.143
4. National Science Foundation - Where Discoveries Begin. (n.d.). Retrieved July 5, 2020, from https://www.nsf.gov/funding/pgm_summ.jsp?pims_id=503286
5. Khaitan et al.,(2014), Design Techniques and Applications of Cyber Physical Systems: A Survey, IEEE Systems Journal, 2014.
6. C.Alippi (2014), Intelligence for Embedded Systems, Springer Verlag, 2014, 283pp, ISBN 978-3-319-05278-6
7. Ramani, Sanjeev Kaushik, and S. S. Iyengar. "Evolution of sensors leading to smart objects and security issues in IoT." In International Symposium on Sensor Networks, Systems and Security, pp. 125–136. Springer, Cham, 2017.
8. Internet of Things: Complete IoT guide - benefits, risks, cases, trends. (n.d.). Retrieved July 1, 2017, from https://www.i-scoop.eu/internet-of-things-guide/
9. Iot Wearable Devices: Should you be using them? (n.d.). Retrieved August 10, 2020, from https://scallywagandvagabond.com/2019/08/iot-wearable-devices-technology-growth-new-products/
10. Swinhoe, D. (2021, January 08). The 15 biggest data breaches of the 21st century. Retrieved April 05, 2021, from https://www.csoonline.com/article/2130877/the-biggest-data-breaches-of-the-21st-century.html

Chapter 3
IoT Security

All IoT sensors and devices are prone to breach of the tenets of information security (confidentiality, integrity, and availability). A study conducted by a technology institute based in France identified about 38 vulnerabilities in terms of poor encryption, presence of backdoors, and possible unauthorized access in the 123 products whose firmware images they tested. IoT is one such domain wherein one weak link could spell doom to all devices connected to the network [1].

The introduction of sensors in devices has made it easy for the attackers to track the activities performed by the user of the device. The fact that these devices are highly pervasive into most places puts the privacy of the user at great risk. Connected devices also help companies to collect sensitive data and create profiles of people that could be compromised by an attacker.

In IoT, we talk about devices that are connected over the Internet, which opens up new doors for the device being hijacked and utilized for illegal activities by the attackers. This gives a new life to the old concept of botnets that became famous after the introduction of distributed computing. A botnet consists of computers that remotely accesses resources without the knowledge of the owner [2]. IoT being a network of many physical systems that are equipped with IP and MAC addresses and transmit and receive data over the Internet makes it an easy breeding ground for attackers to inflict a widespread network attack.

Figure 3.1 depicts a common scenario where the IoT devices are being used by common consumers or a day-to-day basis and the attack surface that it provides to the attackers and an insight into the catastrophic damage that can be inflicted by compromising the devices and the data that is collected by them.

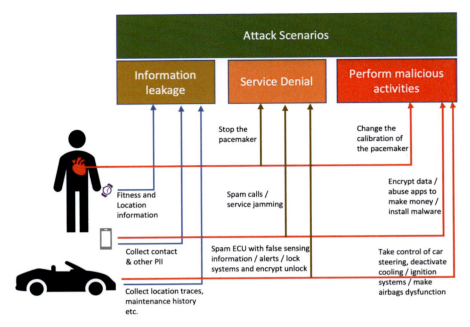

Fig. 3.1 A subset of possible IoT attacks

3.1 Why Do We Need to Secure Smart Devices?

Our connected IoT devices are embedded with one or more sensors and are constantly collecting information making them rich sources of data collection. Even before we come to securing the sensory information (information that is captured through sensors), there is a lot more to the IoT security lifecycle.

3.1.1 Device Bootstrapping

When any new device is to be introduced into a smart environment, it has to be onboarded/bootstrapped into the network. There are various ways to bootstrap these devices, and we will discuss them in upcoming chapters of this book. Failure to bootstrap the devices could lead to the following security issues:

- The device may join a rouge network and start working as a bot in waging attacks like Denial of Service (DoS), Distributed Denial of Service (DDoS), or even scrapping vital and sensitive information and provide it to attackers.
- A rouge device can join a legitimate network and can be controlled by an external attacker to hamper the operations in the network or even bring down the network.

3.1 Why Do We Need to Secure Smart Devices?

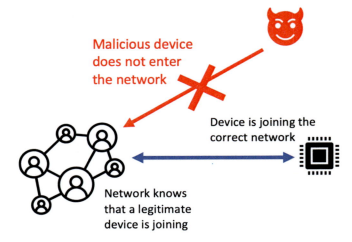

Fig. 3.2 The high-level goal of IoT device bootstrapping

- Possible impersonation or use of misleading identification while interacting with devices in the network.
- The device may not learn/get the required crypto-material about the trust anchor and thus may not even trust legitimate information from trusted sources and thus may be incapable of providing services or working as a part of the system.

Figure 3.2 showcases the purpose of bootstrapping to ensure that a legitimate device joins the network and that a device knows the correct network it should join.

3.1.2 Authentication

After a successful bootstrapping, the device will have joined (registered with the controller) the correct network it has to function/operate in and also have a set of unique cryptographic identifiers that it can use in its communication for the other devices in the network to be able to identify the device in the network. The adoption of IoT and its advocated concept of "everything connected" has provided more ways to access the information these devices hold. Thus, there is a need for encrypting and signing the information that is transmitted by these devices to ensure data integrity and authenticity while also enabling better access control policies.

The process of bootstrapping enables the devices to have their set of keys and certificates that can be used in the information exchange. A common concern among IoT device manufacturers and the application developers is the cost involved in the key generation, signing, and verification processes. The devices being highly resource constrained make it an even larger challenge. With the lack of a standard for securing IoT communication, the application developers have traditionally resorted to not enforcing the need for cryptographic signing or encryption or even the use of multiple keys. This has pushed us into a stage where the devices are ever vulnerable to attack and increased the attack surface enormously.

On the other hand, with the many devices and the need for multiple keys for each of these devices, as a common trend, device owners have been known to store the cryto-material in a single device (like a smartphone) for the ease of use making it a very attractive device to be attacked. Compromise of such a device can open the doors for a multitude of different attacks. Also, as end-users (who are the weakest link in any cyber-system), it is very common for the devices to be used with the generic password/pin that comes with the device and not change it to a stronger, less vulnerable passcode. There are a large number of open-source and free software that can detect the presence of an unsecure system (using a generic password) and provide details to use it as a compromised node. A blog that discusses the step-by-step instructions [3] and the way in which a software called Shodan [4] can be used for taking over non-secured devices like cameras, printers etc. that are not secured. This shows that it is not necessary for the attacker to be well versed but even a script-kiddie (an amateur hacker) can easily get access to a network and cause damage.

3.1.3 Access Control

The other important aspect of IoT security is to identify and restrict who has access to what part of the information that is either collected or processed. A connected environment like a smart-home requires multiple devices to have access to information collected by other devices to be able to automatically perform actuation and thus fulfill the task it was designed for. An example would be the HVAC system: the thermostat in the air conditioning system will have to coordinate with the smoke detectors and alarm system which in turn will have to interact with the flow control system that control the flow of water to the sprinklers in order to turn them on in case there is a fire in the house.

A similar example is the technology of the present and the future is the connected autonomously driving vehicles. These vehicles are heavily embedded with sensors and monitoring systems that monitor the surroundings, the environment within, and thus is able to communicate with other vehicles in the vicinity in being able to maneuver through traffic and reach the destination. Each of the sensors is responsible for a task and malfunctioning of any of these sensors if not rectified by other information fusion means [5, 6] may have devastating outcomes. However, we cannot trust any other vehicle as it is, nor can we offer or accept services from the other vehicles. An important aspect to be considered is to check who has access to what information and thus what are the capabilities it provides them with.

Traditionally, access control has been performed using access control lists (ACLs) and other techniques which do not work well in a highly dynamic and ad hoc environment that the IoT devices work in. In case the devices are moving, there is a possibility of intermittent connectivity making the access control policies even difficult to be designed. However, a robust access control policy is the need of the hour in order to prevent the smart devices and sensors from falling prey to attackers who can use them as bots to wage attacks.

3.2 Challenges to Securing Smart Devices

IoT devices are smart, capable of collecting and generating large volumes of data, etc. but are still highly resource constrained for power, computational, and storage capabilities. Some of the devices are mere sensors that have a limited lifetime determined by the battery that powers it and with very minimal interfaces that provide it with the capability to merely communicate/push information. As an example, consider a flow sensor attached to the water pipes beneath the walls that detects the flow of water and the possibility of any leaks/ruptures to the pipes. Being beneath the walls reduces the interfaces for its communication to either medium aided (vibrations) or other wireless protocols. Expecting these devices to be able to generate keys, request for certificates, encrypt information, sign the information, etc. takes a major toll on their resources as all these operations come at a cost that is not very trivial. In order to extend the lifetime of these resources, the product designers and application developers tend to overlook security and focus more on enhancing the functionality as that would attract consumers who are in need of the functionality and unaware of the security consequences that could prevail.

The other major challenge is the interoperability issues that have come to the fore with the current Internet Protocol (IP)-based Internet being incapable of solving all the issues. Not all the IoT devices are capable of directly connecting to the Internet. Some of them use other wireless communication protocols like Bluetooth, ZWave, LoRA, etc. based on their capabilities and the application they have been designed for. This has been a major concern even in the large-scale adoption of IoT devices in the world.

Along with these concerns, usability is a major challenge to secure the IoT devices. Securing the IoT device in the current scenario requires the user to manually perform bootstrapping, modify the passwords/pins, register keys with a controller, request for certificates that can be used with a device, and many more. An amateur user may not be capable of performing most or all of the above operations. In most cases, thus, the device as bought from the market is just installed in the house and let to communicate with the other devices within and outside the environment making it a potential landmine for attacks to the house. Compromise of one such unprotected device can lead to all the devices being vulnerable, and thus the security of the entire system is jeopardized.

3.3 Can IoT Devices Be Used to Wage Attacks: The Rise of Botnets

Recently, there has been an increase in the number of IoT botnets, which makes the world vulnerable to a large cyber-attack that could spell doom to mankind. In general, a botnet consists of computers that remotely accesses resources without the knowledge of the owner [2]. IoT is not just a network of dedicated computers but also of many physical systems that are equipped with IP and MAC addresses and transmit and receive data over the Internet making it a widespread network that could have a security threat.

Computers that have been compromised and are being used for coordinated attacks and illicit purposes are called zombies or bots. Technically, a bot refers to a device infected by a malware and becomes part of a network of infected devices that would be controlled by a single or group of attackers to inflict an attack. The impact of such an attack could be very huge as once in the net, the bots try to find vulnerabilities in surrounding networks and spread rapidly as Trojan horses and thus infecting a lot of machines [2]. Early records of botnets and studies reveal that they do not consume a lot of network traffic or bandwidth and hence usually go under the radar and are not identified. There have been various malwares that have been translated as a botnet starting with Zeus, which was first detected in 2007. Zeus also known as Zbot has been used extensively to extract financial data and banking information and was also involved in the recent ransomware attack. There are many more such bots in the Internet that have been discovered and are closely being monitored.

One of the early instances of bots being discovered with respect to IoT was reported by a researcher at Proofpoint [2] when it was noticed that maliciously there were thousands of emails logged in a security gateway and originated not just from computers but also from some household appliances like refrigerators. The discovery of such botnets is becoming very common nowadays.

IoT brought about the advent of a market that offers cheap and inexpensive devices like webcams, CCTVs, baby monitors, home security systems, wearables, and other devices that connect to the Internet but do not usually have a well thought about security system in place. Some may not even be password protected. These offer an apt breeding ground for the bots to spread and exploit the conditions. The attacks could range from DDOS to even the encryption of data as in ransomware [7].

A recent study by Forbes magazine reveals that in October of 2016, a botnet was identified comprising almost 100 thousand IoT devices that were not secured and led to loss in availability of an Internet infrastructure provider leading to top websites not being available for a short duration [6]. The other worrying factor is that the botnets utilize all the resources it can harness, and thus there are multiple IPs in use (since each device in the terms of IoT has a unique IP of itself), and it is highly impossible to track or prevent further damage.

Recent researches have shown the use of "Hajime," which is termed as a vigilante IoT worm to clean up the issues spread by illicit botnets (especially "Mirai"—a botnet threatened to take over the Internet world [8]) according to its creator. However, the security of such worms is also a concern as it could easily be abused by attackers for their own misleading ideas. According to Kaspersky labs, Hajime and Mirai exploit the default credential combinations to induce a trial and error way of getting into unsecured networks using the open Telnet ports. This was first identified by people at Rapidity Networks in October of 2016. In the current scenario, the Hajime botnet has a shortcoming that it does not possess any specific implementation of code that can attack a network but is designed only to propagate within the network. Another interesting aspect of the Hajime botnet which seems favorable to the creator's claim that it is safe is that once it gains access to an unsecure IoT device, it blocks access to four ports viz., 23, 7547, 5555, and 5358, which are some of the ports that illicit botnets like Mirai exploit to gain access and cause havoc [8].

3.4 Techniques to Protect Devices

Unsecure IoT devices are potential threats to our safety. There have been instances where botnets as described in the previous section have taken over baby monitors or have even brought the Internet to a standstill. Thus, it is imperative for these devices and their communication modes to be secured so as to protect our own digital lives and safety. A list of possible ways to protect our IoT devices and thus ourselves is discussed below. This list is not comprehensive but a suggestion to improve the current state-of-the-art.

- Use software from trusted sources in your devices ranging from wearables, smartphones, and tablets to even our legacy computers all of which will be a part of the IoT ecosystem. Use of antivirus and anti-malware in supporting devices can help prevent obvious threats through well-known malware.
- Change the generic passcode/pin that comes with the devices to a strong pin/passphrase that is unique and complex. Try to reset the passwords periodically and definitely if there is a sense that the devices are compromised.
- Understanding the privacy policies of the applications that come with the devices and the permissions it requires is critical. Overlooking these policies can prove to be hazardous.
- Identify the data stores of the devices and the links it has to the manufacturer. Check the device for any kind of malicious backdoors or activity that can possibly be a leak of information.
- Try to localize the access of the devices within the network using VPNs so as to restrict public access and thus reducing the attack surface.
- Secure your smart controller. If the controller is a smartphone, then ensure that the device is secured. Compromise of a controller through the social links or other means can lead to the whole system being compromised and an easy source for attack.

3.5 Concluding Remarks

The Internet-of-Things paradigm has changed the way in which we perceive life and has provided us with unimaginable opportunities. However, there are many basic ground-level aspects of the devices and connections that have to be secured but been traditionally overlooked making the sensory nodes that collect information easy targets for the attackers. In this chapter, we discuss the need for bootstrapping, authentication, authorization, access control techniques, etc. that have to be employed with these devices. We also discuss the current challenges to widespread adoption of security in the IoT environment and the possible ways to secure the IoT systems.

In the upcoming unit, we delve into more details related to securing the IoT systems and communication and hence create a connected ecosystem devoid of attacks or hassles for information exchange and decision-making.

References

1. Samani, R. (2016, June 06). 3 key security challenges for the Internet of Things. Retrieved June 26, 2017, from https://securingtomorrow.mcafee.com/business/3-key-security-challenges-internet-things/
2. What is IoT botnet (Internet of Things botnet)? - Definition from WhatIs.com. (n.d.). Retrieved June 25, 2017, from http://internetofthingsagenda.techtarget.com/definition/IoT-botnet-Internet-of-Things-botnet
3. Kody, Ghost, Thetall, Z., Neo555, Dragonhunt3r, Occupytheweb, . . . F, A. (2019, August 07). How to find vulnerable webcams across the globe using shodan. Retrieved April 05, 2021, from https://null-byte.wonderhowto.com/how-to/find-vulnerable-webcams-across-globe-using-shodan-0154830/
4. The search engine for the Internet of Things. (n.d.). Retrieved April 05, 2021, from https://www.shodan.io/
5. Iyengar, Sitharama S., Sanjeev Kaushik Ramani, and Buke Ao. "Fusion of the Brooks–Iyengar algorithm and blockchain in decentralization of the data-source." Journal of Sensor and Actuator Networks 8, no. 1 (2019): 17.
6. Brooks, Richard R., and Sundararaja S. Iyengar. Multi-sensor fusion: fundamentals and applications with software. Prentice-Hall, Inc., 1998.
7. Marr, B. (2017, March 07). Botnets: The Dangerous Side Effects of the Internet of Things. Retrieved June 25, 2017, from https://www.forbes.com/sites/bernardmarr/2017/03/07/botnets-the-dangerous-side-effects-of-the-internet-of-things/#2766bb833304
8. Khandelwal, S. (2017, April 27). Hajime. Retrieved June 25, 2017, from http://thehackernews.com/2017/04/vigilante-hacker-iot-botnet_26.html

Part II
Securing the Internet-of-Things Enabled Smart Environments

Chapter 4
Understanding the Smart Systems: IoT Protocol Stack

The idea of a connected ecosystem with devices being capable to communicate and automatically work toward specified application goals has been around for a long time. In the 1970s, this concept was categorized as "ubiquitous computing," "pervasive computing," or even more colloquially as "embedded computing." The growth of sensory devices and sensing capabilities gave these concepts a new lease of life, and the term "Internet of Things (IoT)" was coined in the work Procter&Gamble by Kevin Ashton [1] in 1999, making it a paradigm known for more than two decades now.

In its new definition, IoT is conceived as a paradigm wherein smart objects equipped with sensory interfaces can perform sensing, actuation, and certain desired applications. This has revolutionized the services industry and transformed human lives in ways that were never even imagined decades ago. The IoT has grown from the simple interactions between physical and digital worlds [2] to manifest into a whole new world where any conceivable device can be created to interact with the human world using computing and networking capabilities. Most of the applications of IoT devices are linked to collaborative functions that it performs with other devices in the network. IoT is not a single technology; rather, it is an agglomeration of various technologies that work together in tandem [3].

4.1 Introduction

Sensory nodes, actuators, and other such CPS devices fall under the category of IoT devices. Each of these devices has specific roles and is a vital cog in the successful operation of an IoT ecosystem. Sensors collect information that is either processed in-house or transmitted to a processing site which on processing the data triggers events that the actuators use to perform actuation and change the state. The term sensor can represent a minute nanoscale device with a single chip or even a

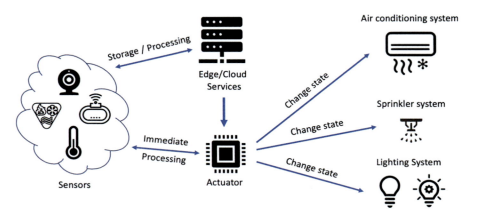

Fig. 4.1 Overall working of the IoT ecosystem. The sensors sense the surroundings and transmit the sensed information to the cloud for persistent storage/processing or to the actuator for immediate processing. The actuators next send the signal to the smart systems that will change state

smartphone that is capable of identifying its state and the state of its surrounding environment in providing inputs for any kind of change in state performed by the actuator. Figure 4.1 depicts a high-level view of the communication among the various IoT devices in fulfilling the needs for a smart ecosystem.

The decision to send the sensory data either to cloud or to a local edge processor for processing depends on the requirement of the application. It is thus a research problem to efficiently allocate the resources and provision the information flow to the appropriate node for the actuator to be able to trigger a state change. Another challenge on which researchers are working on is the ways and means to tolerate ambiguity or failures in the sensory nodes. There are many sensor fusion algorithms and data cleaning algorithms that have been devised to perform adequate preprocessing for effective results.

The other challenges to be addressed are the ones related to communication medium and the protocols used for communication. The communication mode in IoT ecosystems is predominantly wireless using a multitude of communication protocols including but not restricted to Wireless Fidelity (Wi-Fi) [4], Bluetooth (Conventional and Bluetooth Low Energy (BLE) [6]), ZWave [7], Zigbee [8], LoRA [5], etc. The reason for the use of wireless medium is the geographical location of deploying such devices. There are concerns over the use of the traditional IP-based architecture of networking, which has led to the introduction of 6LoWPAN [9], CoAP [10], and other techniques. However, there are still a large number of unsolved research problems.

4.2 Current Internet Protocols: Can They Survive the Meteoric Rise of IoT Device Adoption?

The large-scale adoption of IoT technology has been hindered by:

- How to provision the local and global communication of the devices so as to retrieve or provide services?
- How can data be secured, consistently sent without losses even in a highly lossy environment without any infrastructure support or intermittent connectivity?

The second of the abovementioned points is identified as the basic vision of the IoT technology with devices being able to communicate their state and receive actuation commands to be able to complete the loads of interesting and highly compelling applications. The existing IoT frameworks focus on interconnecting device focusing on the first of the above two points. Figure 4.2 depicts the highly stressed waist (central layer) of the IP protocol in the TCP/IP model of Internet architecture. This networking layer is overworked in the current technologies and is overburdened and responsible for delivering packets from the source to the destination.

The current model, built on this host-centric approach, has showcased issues when working in the IoT communication scenario. The application names used in the application layer cannot be directly used in transmitting and receiving the IP packets. The mismatch in the network layer name and the application layer name has to a large extent restricted the design of modern highly sophisticated applications. The point-to-point emphasis of the communication model works well in situations where we have infrastructure support, but a dynamic environment like that of the IoT applications where there is a lack of constant infrastructure support and intermittent connectivity may overload and burden the Internet which is heavily reliant on the

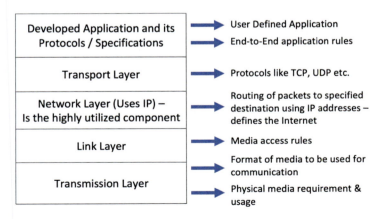

Fig. 4.2 The narrow waist of the current Internet protocol

many mappings that exist in its stack. The mappings make the communication complex and highly brittle.

Consider an example of a smart-light system as the IoT application. Under the current conditions, for a specific light to be turned on or even to modify the intensity of the lights, the control system in the house will have to use the application that will have to get the packets in an appropriate VLAN and IP subnet and use the lighting gateway device's subnet and IP address before dealing with the actual light under consideration. While consumer devices have made this easier, often allowing web-based control over devices on home wireless networks, they do so by making assumptions about such mappings—for example, that all devices are on the same subnet—or rely on cloud services to achieve what is essentially local communication. Thus, there are a lot of issues with the current narrow waist of the Internet, which is an open problem in itself and has led to the use of many IP overlays with the existing networks.

4.3 Available Standards

The IoT stack has two variations: one that is made up of three layers and the other with five layers. Figure 4.3 depicts two different versions of the IoT communication stack as traditionally identified and used in the literature [3]. The three-layer version is the basic architecture with the three layers having the following functions:

- The sensory nodes/devices (capable of perceiving various events and changes in the environment where it is working) form the first layer called the *perception layer*. This layer is often made up of multiple variants of devices, and the data is in most cases aggregated and sent to the next layer for transport and processing needs.
- Following the perception layer is the layer that is responsible for both transporting the sensed data and processing it based on the requirement and design of the application. This layer is called the *network layer* and also plays a vital role in interfacing multiple sensory nodes that are within the perception layer.
- A layer responsible for delivering the specific needs and services for which the application was designed and built is appropriately called the application layer. This layer thus includes the various avenues where the devices can be deployed in and the interfaces that are provided to the humans/controllers to interact with it.

The five layers are perception, transport, processing, application, and business layers. The perception and application layers described below have the same responsibilities as was described in the previous architecture involving three layers.

4.3 Available Standards

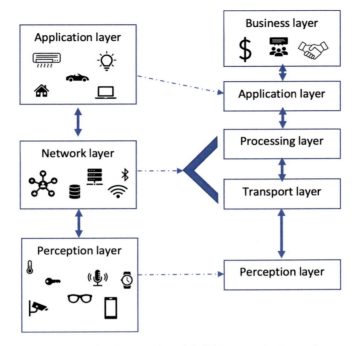

Fig. 4.3 The three-layer and five-layer version of the IoT communication stack

However, the other layers have certain unique features and functions as described below:

- Data collected and aggregated by the sensory nodes are transferred for further processing to the processing layer using the transport layer which is the layer responsible for transferring processed data back to the actuation units as well. The transport layer can use a single or a combination of multiple media and channel options including wireless protocols like Wi-Fi, Bluetooth, cellular technology including 3G, LTE modules, LoRA, RFID, NFC, etc.
- The layer that processes the requests or data received and either provides actuation commands or stores or modifies and sends it to other units/layers for further processing or storage is called the processing layer. This layer is more often also considered as a middleware layer as it functions as a middleman in the entire chain. This layer is involved in data cleaning, fault tolerance, and other aspects of handling various forms and types of data. The processing layer varies in size based on the frequency of query and the amount of data that it usually handles and plays a very vital role in defining the functionality of the entire system. More like the IP layer in the TCP/IP stack, this layer is the backbone and usually is the place that routes and reroutes information to layers below and above as necessary.

- The other layer is entrusted with the overall management and handling of the system and hence is called the *business layer*. It is here that the economic aspects including business plans, monetization structure of the system, payment models, high-level permission definition, and other details are mentioned.

4.4 The Six Layers in IoT Security

The frequency of cyber attacks in the IoT environment is increasing and becoming a major concern. On a high level, the various points at which security could be a concern while using IoT devices are as depicted in Fig. 4.4. Details of these layers and the possible security concerns are described in the below subsections.

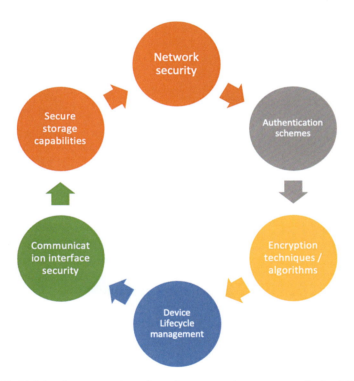

Fig. 4.4 The high-level entrance points of security concerns in a system employing IoT devices

4.4.1 Securing the Network of Operation

Historically, network security was the domain of well-trained experts. However, you also need to understand the basics of network security as you are the administrator of your home IoT network. The first thing you need to remember is not to rush through your IoT setup as it may lead to numerous security problems later. So, be patient. Set up and secure one device at a time.

Unfortunately, most IoT networks are wireless. Securing a wireless network can be a bit challenging as there are several different communication protocols in IoT, standards, and device capabilities. However, one of the easiest ways to protect a network is to change the default usernames and passwords of the broadband router or another wireless access point.

This access point device usually comes with an embedded server or webpage. As the administrator, you can change the username and password here to keep your network secure. You should also change the default Service Set Identifier (SSID). A default SSID often indicates a poorly configured network, increasing the chances of an attack.

4.4.2 Authentication Requirements of Devices

The authentication process keeps unauthorized persons from gaining access to your network while allows you (the administrator) to access the sources you need. You will need to consider different factors to set up necessary authentication rules. For example, you may want to allow multiple users to access a particular device such as your microwave, Wi-Fi, or TV. They may also want to set different parameters according to their needs.

A simple authentication involves providing the users with a username and a password to access a particular device. However, you can also use advanced methods such as two-factor authentication and biometrics. In case of a two-factor authentication, after entering a username and a password, users receive a One-Time Password (OTP) via e-mail or SMS. Alternatively, you can also use digital signatures, personal identification number (PIN), and smart card. Multilayered authentication adds extra protection. Hence, it is preferred in high-security environments.

4.4.3 Encrypting Data to Prevent Leakage of Information in Clear Text

Security of moving and stored data is also a crucial part of overall IoT security. Hackers have known to sniff the moving data to gain illicit access to IoT networks. The high heterogeneity of the devices and the various forms and modes of data that each of these devices can assimilate and use make it difficult to design a single encryption scheme that can fit the needs of all of these.

Most Wi-Fi home networks support encryption technologies like the WPA and WPA2 along with other WEP and other variants of these. Unfortunately, they are not the strongest set of encryption algorithms and have well-defined attack models and threats [11]. The common points of attack include the key length or the number of bits in the encryption key which if not long enough can easily be broken by simple brute force attacks. Encryption keys also suffer the problem of having a predetermined lifecycle because of the possibility of being computed beyond this lifetime by an average attacker equipped with common attack strategies and resources. So, they become worthless after the desired period of usage. Thus, encrypting IoT data also requires efficient encryption keys and key management techniques.

4.4.4 Storage Solutions

Cloud is one of the primary sources of storage and a potential location for cyber attacks capable of destroying the IoT network. Cloud is often considered as the backbone of the IoT setup that connects the smart devices (logically through data) and a central hub where data is analyzed and stored. In other words, it is the big data that requires protection against cyber-attacks.

Most critical security components of the cloud include stored data, platform, and application integrity verification. By default, your service provider is the first line of defense against cloud-based data breaches. They are supposed to provide you with a secure cloud environment. So, choose your cloud service provider carefully. Preferably someone who uses advanced security practices and controls.

4.4.5 Device Lifecycle Management

Often overlooked, device lifecycle management involves keeping all your IoT devices and systems updated regularly. The best way to ensure that everything including operating systems, firmware, and application software remains up-to-date is to turn on automatic updates.

Cyber-attackers continually keep changing their tactics to find new ways to invade secured IoT networks. Most device manufacturers and cloud service providers create security patches to deal with them. So, make sure that the security patches are installed correctly. Follow the required security protocols when adding new devices to the network, end-of-life device decommissioning and integrating your network with a new cloud system.

4.4.6 *Interface or API Protection*

Usually, an application programming interface or API is used to access the devices connected to the IoT setup. API security is critical to ensure that only authorized devices, developers, and apps are communicating with one another. In other words, it maintains the integrity of the data.

You can use a comprehensive API management tool. Most tools can automate connections between an API and the applications. They can also ensure consistency if you are using different variants of API. They can improve performance by managing the memory and caching mechanism as well. Just make sure to select a tool that suits your IoT network requirements.

4.5 Concluding Remarks

In this section, we identified the various versions of IoT protocol stack and their features with functions. This is an attempt to introduce the generic and minimum required layers and is not an extensive list. As the IoT devices evolve and newer versions and capabilities are introduced, there could arise a need for the addition of many other layers or sub-layers that can support such devices. The important aspect from the view of this book is the versatility of these devices and hence the difficulty in designing security protocols and features that can be used in all applications. A one size fits all variant for securing these devices is the desire of security enthusiasts and businesses but still seems to be an area that requires extensive exploration and research.

The next chapter describes in more detail the various stages of the IoT device lifecycle and the security requirements in each of them along with the existing solutions and challenges that are yet to be addressed. We believe the tamper evident security protocol that we propose in this book as an ensemble of the best practices of various techniques will prove to be a compelling technique that can be considered for addressing the security challenges.

References

1. Ashton, Kevin. "That 'Internet of Things' thing." RFID journal 22, no. 7 (2009): 97–114.
2. Vermesan, Ovidiu, and Peter Friess, eds. Internet of Things: converging technologies for smart environments and integrated ecosystems. River publishers, 2013.
3. Sethi, Pallavi, and Smruti R. Sarangi. "Internet of Things: architectures, protocols, and applications." Journal of Electrical and Computer Engineering 2017 (2017).
4. Who we are. (n.d.). Retrieved April 07, 2021, from https://www.wi-fi.org/who-we-are
5. Homepage. (2021, March 03). Retrieved April 01, 2021, from https://lora-alliance.org/
6. Intro to Bluetooth low energy. (2021, April 07). Retrieved April 07, 2021, from https://www.bluetooth.com/bluetooth-resources/intro-to-bluetooth-low-energy/

7. The Internet of Things is powered by Z-Wave. (n.d.). Retrieved April 07, 2021, from https://z-wavealliance.org/
8. The Zigbee Alliance homepage. (2021, March 26). Retrieved April 07, 2021, from https://zigbeealliance.org/
9. Gartner_Inc. (n.d.). Definition of 6LOWPAN - Gartner information Technology glossary. Retrieved April 07, 2021, from https://www.gartner.com/en/information-technology/glossary/6lowpan
10. CoAP - RFC 7252 Constrained Application Protocol. (n.d.). Retrieved April 07, 2021, from https://coap.technology/
11. WiFi security: Wep, WPA, WPA2 and their differences. (2019, October 03). Retrieved April 07, 2021, from https://www.netspotapp.com/wifi-encryption-and-security.html

Chapter 5
Onboarding New IoT Devices

Trust plays an integral part in network communications. In an IoT environment with multiple communicating entities, the need for trust is more pronounced. The presence of trust with the communicating entity ensures the services or updates the devices are willing to accept between them. In a broad sense, trust can be categorized as being *static* (based on pre-acquired knowledge) or *dynamic* (based on current interactions). In the case of IoT devices or networks that rely on sensory communications, static trust establishment is not feasible always, and there is an inclination to the use of dynamic trust.

In traditional approaches to trust establishment, the communicating parties can rely on preexisting configurations; preconfigured sets of root Certification Authority (CA) certificates and transitive trust in Public Key Infrastructure (PKI) model—or dynamically build trust relations—using feedback or explicitly setting trust decisions of certificate trustworthiness in Web-of-Trust (WoT) model. In a highly dynamic IoT environment involving motion (like vehicular networking), these traditional methods may not work. PKI and WoT usually require connectivity to infrastructure which may not be feasible to maintain due to the mobility patterns of the vehicles. The trust relationships thus defined are usually long-standing and not applicable to mobile IoT applications.

5.1 Talking to the Correct Device/Network?: The Need for Device Bootstrapping

Any new service or interaction with providers of the service needs initialization of trust as the first step before delving into deeper trust relationships. Bootstrapping can be defined as an onboarding process through which an entity learns the presence of other entities in the network along with learning of the other entity's current state. It is the process that prevents the infiltration of malicious nodes into the network and helps in making sure that legitimate devices or entities are getting the necessary accesses or privileges. Trusted device bootstrapping is thus a technique that aids in the requestor of services making the correct choices and through the correct means [1]. Large-scale proliferation of sensors in the world and the evolution of sensors [2] have led to the automation of the bootstrapping process. However, in an IoT ecosystem, there are significant limitations that have to be overcome based on the device types and the types of guarantees that will have to be offered against attackers.

The initial stages of onboarding (bootstrapping) indicates the need for the devices (or device and controller) to exchange cryptographic information that can be the initial facilitator of communication. On a high level, the overall goal of the bootstrapping phase of an IoT device's lifecycle is to ensure that the device eventually joins the *correct* network where it can operate and the network is aware of the device that is joining and is able to determine the *legitimacy* of the device.

The entire process of bootstrapping is usually highly cumbersome. In an IoT environment, the challenges are even pronounced as the devices are highly heterogeneous. The devices are designed to work with minimal resources in terms of interfaces available for information exchange/communication media, power, location of operation, and other constraints that make this operation difficult. To add to the concerns, the primary mode of communication is usually chaotic and filled with attackers who are constantly listening and awaiting information that can be used to compromise the system. This makes these interfaces highly vulnerable and hence unusable for transmitting the important bootstrapping information. To accommodate this issue and provide a means for effectively completing the onboarding of devices, we use alternate channels also often called as out-of-band or auxiliary channels.

With the existing IoT devices and their capabilities in terms of communication, there are various modalities of the OOB communication/channels. The common options available are highlighted in Fig. 5.1 and can be categorized in broad aspects

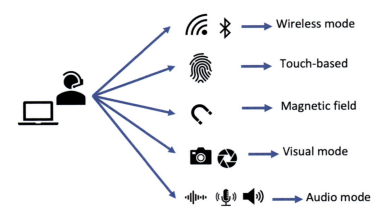

Fig. 5.1 Existing out-of-band communication methods

to be either using wireless media that includes Wi-Fi [3], Bluetooth, Bluetooth Low Energy (BLE) [4]; touch-based/haptics in fingerprint sensors, multi-touch patterns, soft buttons, etc.; image/visual options that include cameras, iris/retinal scanners, image recognition tools, QR code or barcodes, gesture patterns, etc.; electromagnetic field signatures of devices that involve the use of magnetometers, heat sensors, etc.; audio-enabled approaches like acoustic wave patterns, vibrations, SONARs, RADARs, LIDARs, ultrasonic sensors and receivers, etc. to name a few. However, it is not feasible to design a generic bootstrapping technique that can be applicable for all devices as not two devices are similar or have the same capabilities. Building multiple bootstrapping schemes for individual devices can be cumbersome and an overkill in most cases. So it is important to understand the capabilities of devices and the various application scenarios they will be involved in while considering the bootstrapping technique to be used.

Allowing a malicious device into the network can lead to catastrophic damages depending on the application and environment of operation of the devices and network. Some of the abovementioned options lead to the creation of static techniques that can further have security implications making this stage of the IoT device lifecycle very important. In some cases, there could be a need to bootstrap multiple devices at once or to form a new group within the network, and the research in this field is still in its nascency.

5.2 Taxonomy of Out-of-Band (OOB) Approaches

Out-of-band (OOB) approaches are characterized by the use of non-conventional channels or auxiliary frequencies that are highly contrasting to the specific mode of communication that the device is designed to operate. The main motive for the use of these OOB channels is for the exchange of bootstrapping information that includes cryptographic keys and material that can effectively improve and enhance the system's security. Figure 5.1 depicts a subset of the various modes/channels that can be explored to be used as the OOB channel in an IoT ecosystem and thus for the bootstrapping/re-bootstrapping of the devices in their operating environment. This section surveys some of the existing techniques that can be modeled to be used for bootstrapping. We also give a brief insight into the security properties of each of these techniques so that a best fit can be identified based on the specific application, capabilities, and operating environment of the devices.

5.2.1 Bluetooth Low Energy (BLE)

Bluetooth Low Energy (BLE) has over the past decade become a common choice of communication in IoT networks and is also a good choice to be used for OOB communication. An example of the use of BLE in an OOB technique for exchange of bootstrapping information is discussed in the design of Device Provisioning Protocol (DPP) [5]. In general, such techniques require (a) a device (usually the controller) that configures the environment and broadcasts a signal which is considered as the intent and (b) a device that is to be onboarded assuming the role of an enrollee. The enrollee is entrusted to respond to the broadcasted intent when it enters the range of communication.

After such an initial exchange is complete, an auxiliary channel defined by BLE will be used to exchange the cryptographic information, which includes keys and information related to the environment's trust anchor. BLE is usually easy to adopt and use and hence is considered as a highly viable option and many modern IoT devices already come enabled with Bluetooth. However, there is a major drawback of using BLE-based approaches because the pairing weakens as the devices start drifting apart or reach the extremes of the communication range opening the possibility of eavesdropping and hence information leakage. The BLE protocol inherently does not hold on to the bootstrapping keys during the transfer, and hence there could be losses leading to corrupted exchange of bootstrapping information and the lack of proof of possession makes it easy to repudiate.

5.2.2 Haptics/Touch

Another common approach to bootstrapping devices that have an interface that has a touch-enabled screen or a trackpad is the use of haptics. Even physical button-based approaches where a specific pattern can be punched in can be considered in this category. Security designers can use this technique specify patterns and use them for the initial trust establishment thus giving the device information that it should know about the network it will be operating in. The entering of a password or even use of biometrics falls under this category of OOB channels.

An example of how a specified pattern of button presses can be used for bootstrapping or even exchange of information is described in the button-enabled device association (BEDA) protocol [6]. The pattern generated by the button presses is usually translated into information that can be decoded by the device to understand and identify the required bootstrapping information. It is argued that unless there is a camera or any visual aid that can capture multiple frames a second and thus be able to replay the pattern, it is difficult to replicate it. However, the drawback is the reliance on physical capability of a human (in most cases), which leads to errors being induced in the system. The sensitivity of human touch especially on physical button presses can vary tremendously, and hence the fault tolerance of the system has to be high and at times can lead to even legitimate devices to be rejected by the network. More importantly, not all IoT devices (especially ones that an extremely small in size and resource constrained) can afford to have a touch-enabled interface, and hence this approach is not viable for all scenarios.

5.2.3 Magnetic Field Technique

All IoT devices consume power and hence usually have an electromagnetic signature surrounding it. The magnetic field intensity and orientation is argued as a viable option for exchange of information. Jin et al. [7] in their paper discuss the use of such a technique for pairing smartphones. According to their design, a magnetometer is used to identify and record the device's surrounding magnetic intensity along with which additional attributes like device orientation and ambient noise are recorded and used to encode the required cryptographic information and thus used for bootstrapping. In order to neutralize the effect of possible eavesdroppers, the authors have suggested the use of a specialized *interlock* protocol that also aids in preventing certain passive attacks. The protocol is also argued to be tolerant to Denial of Service (DoS) attacks because it is very difficult to overload the magnetometer reading without the influence of an extremely large magnet.

The major drawback of this method is the complexity and difficulty involved in the generation of stable signals and thus the manipulation of resulting magnetic fields. Also, for precise encoding of information and subsequent transmission of the information, there is a need for bulky coils and other resources that are not usually available in an IoT environment.

5.2.4 Visual Techniques

Visual techniques are a broad category of OOB channels, which includes cameras, screens, light sensors, photodiodes, proximity sensors, and many more. An example of how visible light can be altered to be used for onboarding devices is discussed by Kovacevic et al., in [8]. In the specific design discussed, the device uses photodiodes as photoreceptors to detect and use combination of light pulses as the encoded message sent from a controller. A naive but major vulnerability of this approach is that it is easy for an onlooker to identify the pattern and hence be able to replicate it or even decode it to identify the cryptographic information and hence work toward an attack model against the system. There is also the need for specialized photo sensors that are not very common among the current crop of IoT devices.

5.2.5 Audio Techniques

A specialized pattern of sonic frequencies has also proven to be a possible technique for onboarding devices. One such approach is explained by Soriente et al. [9], where information is exchanged using specially devised codecs that are capable of generating audio that would be gibberish to humans. Such techniques though are highly prone to DoS attacks as it has become common to produce devices that can cause destructive interference and thus jam the network or even corrupt the data being transmitted.

Ultrasonic frequencies though not in the human audible range are also an acoustic technique that is commonly used for these purposes. Mayrhofer and Gellersen in [10] describe a use case for this technique. A major drawback of this approach is the need for highly controlled surroundings for effective pairing. A common form of attack involves the tampering of connectivity by altering device locations to remove them from direct line-of-sight enabling information leakage or interruption using a man-in-the-middle (MITM) attack.

5.2.6 Vibration Techniques

Prior attempts to use vibration as a mode of communication for security purposes can be seen in [11–15]. Lee et al. [16] describe an approach to enhancing the communication rate when using a vibration channel to pair devices. These articles discuss the benefits of the use of vibration for secure communication and its role in alleviating the threats of other techniques. Controlled vibrations can be effectively used if the devices are very close to each other, thus creating a reduced attack surface. However, there are trade-offs related to the data rate and the bit errors that can occur due to lack of synchronization.

5.3 Existing Bootstrapping Techniques

Trust bootstrapping as defined earlier is a mechanism of assigning trust rates for new devices and services in a network and thus compute the trustworthiness of the entity. This process is considered a part of the trust-building phase and is performed among entities with little or no prior interactions. Earlier works in this field approached this process by assigning default values which will be altered later to either increase or decrease based on the model and criteria introduced in [17]. The concept of community-based bootstrapping is discussed in [18]. The dependence of the community-based approaches is one of the limitations of this work. A user should be able to trust a service before its invocation without requiring the existence of a community that evaluated the service in the past.

5.4 Concluding Remarks

Bootstrapping is the first step of the IoT device lifecycle and has to be performed in order to ensure the correct device joins the networks and adversaries are kept at bay. After successful bootstrapping, there is usually a mutual trust among the devices in a network and that can further be enhanced based on the activities performed by them in the network in due course of time.

To emphasize the importance of this stage in the entire device and network lifecycle, the authors have written articles that describe a vibration-assisted automated bootstrapping technique [19] and a technique to induce rapid trust in highly dynamic environments like that of vehicular networks [20], both of which use a new data centric approach of communication called Named Data Networking. In the next unit, we will briefly introduce the history of cryptography and how it is used as an integral part of security and build on the techniques that motivated us toward building a tamper evident design to security.

References

1. Parno, B., McCune, J. M., and Perrig, A. Bootstrapping trust in commodity computers. In 2010 IEEE Symposium on Security and Privacy (2010), IEEE, pp. 414–429.
2. Ramani, S. K., and Iyengar, S. Evolution of sensors leading to smart objects and security issues in IoT. In International Symposium on Sensor Networks, Systems and Security (2017), Springer, pp. 125–136.
3. Who we are. (n.d.). Retrieved April 07, 2021, from https://www.wi-fi.org/who-we-are
4. Intro to Bluetooth low energy. (2021, April 07). Retrieved April 07, 2021, from https://www.bluetooth.com/bluetooth-resources/intro-to-bluetooth-low-energy/
5. Wi-Fi Alliance – Device Provisioning Protocol, "Device provisioning protocol specification v1.1," 2018.
6. C. Soriente, G. Tsudik, and E. Uzun, "Beda: Button-enabled device association," 2007.
7. R. Jin, L. Shi, K. Zeng, A. Pande, and P. Mohapatra, "Magpairing: Pairing smartphones in close proximity using magnetometers," IEEE Transactions on Information Forensics and Security, vol. 11, no. 6, pp. 1306–1320, 2015.
8. T. Kovačevic, T. Perkovic, and M. Čagalj, "Flashing displays: user- friendly solution for bootstrapping secure associations between multiple constrained wireless devices," Security and Communication Networks, vol. 9, no. 10, pp. 1050–1071, 2016.
9. C. Soriente, G. Tsudik, and E. Uzun, "HAPADEP: human-assisted pure audio device pairing," in International Conference on Information Security. Springer, 2008, pp. 385–400.
10. R. Mayrhofer and H. Gellersen, "On the security of ultrasound as out- of-band channel," in 2007 IEEE International Parallel and Distributed Processing Symposium. IEEE, 2007, pp. 1–6.
11. S. A. Anand and N. Saxena, "Vibreaker: Securing vibrational pairing with deliberate acoustic noise," in Proceedings of the 9th ACM Conference on Security & Privacy in Wireless and Mobile Networks, 2016, pp. 103–108.
12. Y. Kim, W. S. Lee, V. Raghunathan, N. K. Jha, and A. Raghunathan, "Vibration-based secure side channel for medical devices," in 2015 52nd ACM/EDAC/IEEE Design Automation Conference (DAC). IEEE, 2015, pp. 1–6.
13. N. Saxena, M. B. Uddin, J. Voris, and N. Asokan, "Vibrate-to-unlock: Mobile phone assisted user authentication to multiple personal RFID tags," in 2011 IEEE International Conference on Pervasive Computing and Communications (PerCom). IEEE, 2011, pp. 181–188.
14. A. De Luca, E. Von Zezschwitz, and H. Hußmann, "Vibrapass: secure authentication based on shared lies," in Proceedings of the SIGCHI conference on human factors in computing systems, 2009, pp. 913–916.

15. R. Kainda, I. Flechais, and A. Roscoe, "Usability and security of out- of-band channels in secure device pairing protocols," in Proceedings of the 5th Symposium on Usable Privacy and Security, 2009, pp. 1–12.
16. K. Lee, V. Raghunathan, A. Raghunathan, and Y. Kim, "Syncvibe: Fast and secure device pairing through physical vibration on commodity smartphones," in 2018 IEEE 36th International Conference on Computer Design (ICCD). IEEE, 2018, pp. 234–241.
17. A. Rowstron and G. Pau, "Characteristics of a vehicular network," University of California Los Angeles, Computer Science Department, Tech. Rep, pp. 09–0017, 2009.
18. A. Compagno, M. Conti, P. Gasti, and G. Tsudik, "Poseidon: Mitigating interest flooding DDoS attacks in named data networking," in Local Computer Networks (LCN), 2013 IEEE 38th Conference on. IEEE, 2013, pp. 630–638.
19. Ramani, Sanjeev Kaushik, Proyash Podder, and Alex Afanasyev. "NDNViber: Vibration-Assisted Automated Bootstrapping of IoT Devices." In 2020 IEEE International Conference on Communications Workshops (ICC Workshops), pp. 1–6. IEEE, 2020.
20. Ramani, Sanjeev Kaushik, and Alex Afanasyev. "Rapid Establishment of Transient Trust for NDN-Based Vehicular Networks." In 2020 IEEE International Conference on Communications Workshops (ICC Workshops), pp. 1–6. IEEE, 2020.

Part III
Cryptosystems: Foundations and the Current State-of-the-Art

Chapter 6
Introduction

6.1 Motivation

Rapid proliferation of highly affordable devices and sensors sparked the age of cyber-physical systems. Humans and devices started seamlessly interacting to give life to the Internet-of-Things (IoT) paradigm. The systems involved in building such smart environments rely heavily on the plethora of information that is sensed, aggregated, processed, manipulated, and leveraged to get the desired services. Information thus has got a new identity and is not static anymore. Traditional storage methods have also been replaced with modern techniques with transferring storage or purchasing storage as a service from cloud providers being the norm today. With our high reliance on data and the use of third-party storage, there are a few important concerns that arise:

- How can we ensure our information is safe in the stored location?
- Who has access to personalized information that we store in such locations?
- Is the integrity of the data maintained or is data being altered?
- Who is liable if the storage is compromised or corrupted?

All the abovementioned questions are of great importance in this cyber era. Greater importance is provided to enhanced processing capabilities than to securing the information. Also, the modern trends to use data in training machine learning models for automating and artificial intelligence (AI) based tasks has exposed information in the open with attackers identifying every possible way to reach out to it and exploit it with malicious intent. The secure transmission and reception of this data and the storage either in the end devices or in a mostly untrusted environments are a great challenge to be addressed.

6.2 Cryptographic Solutions to Secure Information

As a solution to the above discussed issue related to information leakage and loss, evolved the field of cryptography. The simple idea of cryptography is to manipulate the data at the site of production using a specialized algorithm termed as a cryptographic algorithm so that it will not have any value to intruders in the path. *Encryption* is the specialized term used for this operation and is an important aspect in cryptology. Encryption schemes enabled the sender of an information to be assured that the information is not sent in plaintext (human/machine readable language) thus ensuring the confidentiality of information. *Confidentiality* along with *integrity* and *availability* are the three tenets of information security.

However, if the information has been transformed into a new form using encryption techniques, there is also a need for an operation that can retrieve the original message. The techniques involved in the recreation of the initial plaintext is studied under another field called *cryptanalysis* with the specific technique of inferring the message from an encrypted form being referred to as *decryption*.

6.3 History of Cryptography

The history of cryptography is a mystery yet to be unearthed. However, the earliest known use of cryptography dates back to the times of the Roman emperor *Julius Ceasar*. Ceaser is expected to have used a simple modification scheme to encrypt his information exchanges with his generals and thus prevent the enemies from identifying the transmitted information. In his honor, we have the famous substitution cipher named "Ceasar Cipher" which is one of the most primitive symmetric key cryptographic schemes. Taking cues from this scheme, the field of cryptography has evolved leaps and bounds. The forthcoming chapters in this book will discuss more advanced cryptographic schemes highlighting the evolution of the field. It is to be noted that as the security of information was being pondered upon, there rose the crop of people (commonly called "attackers") who maliciously tried to break into the system by devising attacks. Securing information from such entities and thus creating a tamper evident and secure ecosystem is discussed through the length of this book.

Figure 6.1 shows an old note that was exchanged between Queen Mary and Sir Babington. On first sight, the letter may seem to have combinations of strange symbols strung together. But in reality, each of these characters could have meant something more useful for the receiver of the message. These symbols were in fact identified to hold a coded message as identified by specialists and led to the execution of queen Mary subsequently.

6.5 Issues with Existing Techniques

Fig. 6.1 Snippet from encrypted communication between queen Mary and Sir Babington [1]

6.4 Current State-of-the-Art

Advances in communication methods and storage techniques with the rapid generation of information has ignited the need for newer and robust techniques that can secure this information. Research efforts are focused on providing simple, usable techniques to benefit the broad scope of computing users. The compelling need is for a solution that can seamlessly scale and support the exponential growth of computing technology. There is also a necessity to define novel ways of performing secure transactions even in highly untrustworthy environments.

The field of cryptography has grown from strength to strength beginning with the symmetric cryptography to the modern updates to the widely used public key cryptographic techniques. Researchers are trying to revamp the public key infrastructure based techniques. This effort has given a new lease of life to the development of asymmetric cryptosystems that can be used in securing information (Fig. 6.2) and data in multi-party systems. Evolving communication systems have been stretching these cryptosystems to their limits since most organizations and users expect to have secure data exchange with minimum latency even in highly unsecure environments. To add to this, privacy and anonymous exchange without any losses is the expected order of the day.

6.5 Issues with Existing Techniques

The generic backbone infrastructure that most modern cryptosystem designs depend on is based on the Public Key Infrastructure (PKI). However, PKI has its share of challenges and vulnerabilities that are very often exploited by the adversaries who are using novel means to bypass the security or bring systems down. It is to be noted that new frameworks like cloud services and storage to even the most modern quantum-computing procedures use the PKI technique and thus fall vulnerable to the attacks.

Fig. 6.2 Cartoon depicting the use of encryption in the modern day [1]

Let us consider an autonomously driving car that seamlessly navigates and maneuvers through traffic. The luxury of this is made possible because of the sensing, aggregation, and exchange of information at high data rates. The vehicular scenario is a great example of generation of large volumes of information that can reveal a lot about a person if not dealt with appropriately. There can be information as to the trajectory that is traversed every day, the services that are requested for, and to even minute things as to when the vehicle was refueled and for how long it could last which gives adequate information for a malicious person to inflict catastrophic harm to the vehicle or the passengers in it.

The make or break of an economy and the management of a country rests largely on the shoulders of the people running the government. These representatives are usually identified, voted for, and picked by people. Modern systems work with the use of electronic ballots with the help of remote servers and cloud services. Preserving the integrity of the information in such servers is of utmost importance as it can have a telling impact on the future of a nation. Also, the new means adopted by teams into using social-engineering to influence and manipulate the decisions of people makes it a highly difficult environment to safeguard one's information.

The well-known cryptographic technique presented by Rivest, Shamir, and Adelman in the form of RSA provided the first practical and deterministic demonstration of the use and power of PKI [2]. It was a great advancement with the users and entities each having their set of keys that they can safeguard and use to protect their information. Also, the addition of hashing schemes and the emergence of

6.5 Issues with Existing Techniques

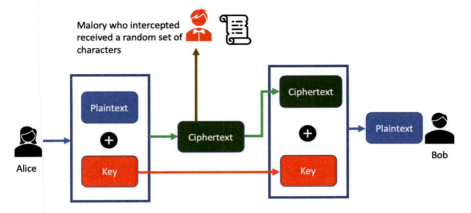

Fig. 6.3 General PKI based encryption scheme

digital signatures have ensured the integrity and non-repudiation of the activities and information. People and entities have been made to take up responsibility and become accountable for their activity in this cyber era with the use of digital certificates.

Improvements on the traditional PKI system (Fig. 6.3) were provided by the addition of probabilistic techniques by Goldwasser and Micali [3] using quadratic residuosity. The addition of probability decreases the threat and allows for risk management and mitigation schemes. However, the work done by researchers on these seminal works has been incremental in improving specific aspects that work for some applications and not all and a complete revamp of the system is required. With the attackers gaining in strength with greater computational abilities and the larger use of technological loopholes, there is an impending need for novel changes and a transition in existing technology to be able to tackle these issues. In this book, we propose crypto-algorithms that can redefine the way in which data is secured and exchanged.

Our major proposition incorporates the lessons learned from the development of the existing cryptosystems and is built based on the best practices that have evolved out of these. Our model is a paradigm shift in the way information security is perceived in this cyber era and tries to address the usability and experience of the consumers by providing better Quality of Service (QoS) and Quality of Experience (QoE) because the user is the entity in the cybersecurity world that is considered an important aspect in the cyber-defense strategies and can be the defining entity in the whole system.

The paradigm shift required in securing information includes a transition/shift of the following kind:

- Safeguarding information governed by specific conditions → Unconditional information security
- Bit level encryption (slower) → encrypting blocks/manifests if possible

- Dependence on certain security parameters → Independence of any external factors
- Security algorithms whose run time is asymptotic → Running time being practical

6.6 Do We Have a Solution?

The general notion of cryptosystems among users is that a message is encrypted to generate a ciphertext which on reception is decrypted to reveal the original message. As an addition to this, modern schemes like the homomorphic encryption scheme can introduce computations to be performed on data even without revealing the actual information (Fig. 6.4).

The computations that can be performed on the ciphertext decide the type of homomorphism. There are many encryption schemes that have been proposed which allow partial homomorphic encryption, i.e., they allow some minimal mathematical manipulations or computations. It was only recently, in the last decade, that the first fully homomorphic encryption scheme was designed by researcher Craig Gentry who describes it as a blackbox in which we can perform operations without exposing it to the external world [4]

On the whole, by using a homomorphic encryption technique, we can process and manipulate the data when it is in its encrypted state. This technique has led to its large-scale adoption in cloud based services ensuring that either the cloud service provider or any third party can only apply functions and computations as a part of the exchange or storage process without having to actually reveal the contents.

The most common variant of the homomorphic cryptosystem is based on the asymmetric PKI model while there are instances of symmetric key systems as well. The most common operations that are performed on the ciphertexts are addition and multiplication allowing the encrypted data to be manipulated and analyzed as though it is in plaintext format without actually decrypting it. The users can compute and process the encrypted data to get an encrypted answer, but only the rightful owner with the designated private key can decrypt to reveal the original message [5].

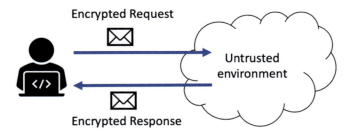

Fig. 6.4 Homomorphic encryption

Certain cryptographic algorithms exhibit the property of malleability wherein, it becomes possible to transform a generated ciphertext into another ciphertext such that it decrypts to a related plaintext. That is, given an encryption of a plaintext m, it is possible to generate another ciphertext which decrypts to $f(m)$, for a known function f, without necessarily knowing or learning m.

The above described malleability property is not desired in many cryptosystems because of the identified vulnerabilities it has in allowing an attacker to alter a message. It provides the attacker with the power to modify the messages even without having any knowledge of the ciphertext. However, the manipulation of the ciphertext using algebraic techniques as defined earlier is a desired property in systems designed to perform homomorphic encryption.

On the other hand, some cryptosystems are malleable by design. In other words, in some circumstances it may be viewed as a feature that anyone can transform an encryption of m into a valid encryption of $f(m)$ for some restricted class of functions f without necessarily learning m. Such schemes are known as homomorphic encryption schemes and have been discussed earlier.

A cryptosystem may be semantically secure against chosen plaintext attacks or even non-adaptive chosen ciphertext attacks (CCA) while still being malleable. However, security against such adaptive chosen ciphertext attacks (CCA) is equivalent to non-malleability.

6.7 Final Remarks

The field of cryptography has evolved over the years with many new variants being proposed and used that help in achieving the three tenets of security in the form of confidentiality, integrity, and availability. As the security of the systems evolves, there are also attacks that have rapidly grown from strength to strength and identifying a way to thwart such attacks is a major challenge that needs a lot of research and implementation. In this book, we propose a tamper evident approach to achieve this security in modern applications like cloud services involving storage and retrieval.

References

1. An early evidence of the use of cryptography. (n.d.). Retrieved from https://intensecrypto.org/public/lec_01_introduction.pdf
2. Milanov, Evgeny. "The RSA algorithm." RSA Laboratories (2009): 1–11.
3. Goldwasser, Shafi, and Silvio Micali. "Probabilistic encryption." Journal of computer and system sciences 28, no. 2 (1984): 270–299.
4. Greenberg, A. (2017, June 03). Hacker lexicon: What is homomorphic encryption? Retrieved April 07, 2021, from https://www.wired.com/2014/11/hacker-lexicon-homomorphic-encryption/
5. Crane, C. (2021, March 12). What is homomorphic encryption? Retrieved April 07, 2021, from https://www.thesslstore.com/blog/what-is-homomorphic-encryption/

Chapter 7
Symmetric Key Cryptography

Symmetric key cryptography (or symmetric encryption) deals with the use of a single key for both the encryption and decryption of messages. It was the most prominent approach to secure information for decades and is still used certain systems with other schemes.

Modern computer systems still highly rely on the variations of Symmetric key encryption (Fig. 7.1) algorithms to ensure data security and user privacy. The Advanced Encryption Standard (AES) and Data Encryption Standard (DES) [3] are the most famous names that are usually associated with symmetric key cryptosystems. Many cloud service providers and messaging applications use symmetric ciphers even to this date. The advantage of embedding schemes into the hardware makes schemes like AES 256 an important player in this field.

7.1 Applications and Advantages

One of the major advantages of the symmetric key cryptosystems is the highly affordable time and space complexity of the associated algorithms. The systems are highly secure too especially with the new crop of AES algorithms. To improve the security properties of a system that deploys symmetric key cryptography, tweaking the key by increasing the length of the key will suffice. Performance analysis on existing schemes has shown an exponential increase in the difficulty of cracking a system encrypted with a key scaled by a single bit.

As discussed earlier, the most common use cases of these algorithms are when it is used along with other asymmetric schemes as a hybrid cryptosystem. A working example of such a hybrid system that most of us use in our day-to-day life is in web-based security. The Transport Layer Security (TLS) protocol that secures our networks and even the Internet on the whole implements a hybrid scheme.

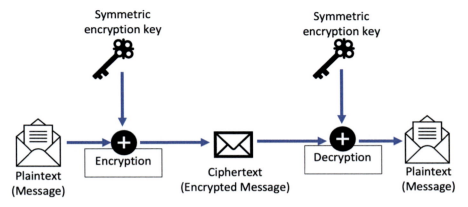

Fig. 7.1 Symmetric key cryptosystem

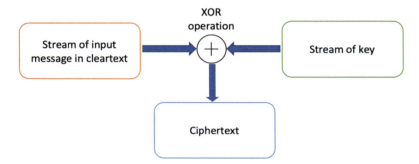

Fig. 7.2 Stream cipher

7.2 Stream Ciphers

One of the most common types of symmetric key cryptography is to stream the data as bits and use an XOR operation with a stream of keys for encryption and decryption. There is usually a bitwise transformation of the input stream and hence the name stream cipher. Figure 7.2 highlights a high-level view of how stream ciphers are designed and used. There are various subtypes within stream ciphers but will not be highlighted in too much detail here.

7.3 Block Ciphers

Similar to the above described stream cipher, block ciphers are a type of symmetric key cryptosystem. One of the major differences between the two of them is that the conversions/operations in this method happen on blocks (chunks of specified

Fig. 7.3 Block cipher

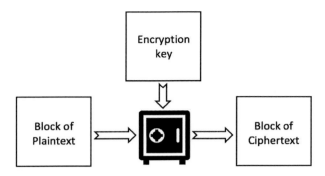

sizes) of data instead of each bit at a time. The block size and the corresponding key lengths lead to the various options available under block ciphers. In some cases, there is an introduction of an Initialization vector or even some kind of randomness or pseudo-randomness which makes the cryptosystem reliable until they can be broken.

AES (Advanced Encryption Standard) [2] and DES (Data Encryption Standard) which have been adopted by the U.S. National Institute of Standards and Technology (NIST) are common variants of block ciphers that are in extensive use even in today's world. Another unique feature of the block ciphers is that even if there are multiple copies of the same block of data, due to the randomness that is introduced, the output of encryption will be a unique block and hence leads to the high reliability. Most crypto-libraries have AES and DES defined that can be directly utilized in our implementations.

Like the stream ciphers, there are various modes of operation of block ciphers. Some of the various modes of operation for block ciphers include CBC (cipher block chaining), CFB (cipher feedback mode), CTR (counter mode), GCM (Galois/Counter Mode), etc. [1]. Figure 7.3 is a high-level view of how Block ciphers are generated.

7.3.1 Initialization Vectors

Symmetric key cryptography is still being used for the economic benefits it provides and the ease of key management as the encryption and decryption use the same keys. And traditionally stream and block ciphers have been a first choice for more internal systems where the communicating entities have some kind of understanding or prior knowledge of the other. To add some randomness and not have a pattern erupt that can make the systems vulnerable, a specialized block of bits called the Initialization Vector (IV) is used along with the key for the encryption and decryption. The reduction in possibility of pattern generation and thus the lower probability of the system compromise reduces the need for re-keying the system which can be very cumbersome in a large organization of communicating entities.

The IVs [4] are used for both stream and block ciphers and thus vary for both these approaches. It could be a direct XOR with a block of bits or the use of multiple iterations to identify the bits that will form the IV. In most cases to ease the decryption process, the receiver of information should be aware of the IV that was used during encryption. There are various techniques to communicate the IV to the receiver. A simple and naive approach is to add the IV along with the ciphertext (not an advisable option) or have other means of understanding from previous communication as to what would be the IV. In some cases, IVs are provided by secure tokens (similar to ones that are used as a part of Multi-factor Authentication (MFA)). It is usually advised to have a unique IV for each cryptographic operation so that there is no overlap or collision leading to a possible compromise of the system.

7.4 Major Disadvantages of Symmetric Key Cryptosystems

All encryption schemes do have some vulnerabilities and have shown to have been compromised numerous times. More often, the reason for compromise is faulty implementation of the cryptographic algorithm which even makes brute force look like an easy option to break into systems and networks.

Though symmetric key cryptographic algorithms can be argued to be simplistic, easy to implement, and even economical, the major disadvantage is the need for properly managing the keys because both the encryption and decryption use the same keys. If the keys are not shared in some secure fashion, there is always a possibility for an adversary to identify the key and probably even eavesdrop on the packet exchange and be able to extract, decrypt, and view the message thus compromising the system.

Thus it is now a common practise for security designers to design the system using a combination of symmetric and asymmetric key cryptography to overcome the abovementioned issue. The modern Internet architecture uses Transport Layer Security (TLS) to secure content being transferred and is a great example of the combination of the symmetric and asymmetric variants.

7.5 Cryptographic Attacks on Symmetric Key Cryptosystems

The most commonly seen attacks on data encrypted using symmetric key cryptographic algorithms fall into three main categories namely:
1. Key search/brute force attacks
2. Cryptanalysis
3. Systems based attacks

7.5 Cryptographic Attacks on Symmetric Key Cryptosystems

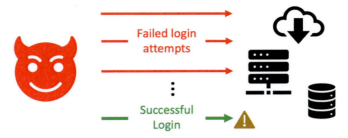

Fig. 7.4 Brute force attempts to gain access to keys

7.5.1 Key Search (Brute Force) Attacks

The most naive of approaches to break an encrypted message is by using brute force. As the name suggests, this technique basically involves the use of every possible key to try to reveal the message and continue the process until the key is found. Thus the smaller the key, the lesser the number of combinations to be tried and the more potent the brute force attack is. Thus the easiest way to protect against this attack is to have a considerably large key that cannot be easily broken even by the modern supercomputers. Another important aspect to keep in mind is that, apart from altering the key and key size, there is no other way to prevent key search based attacks. In most cases, a 128-bit key has shown to take enormous amounts of time (in some cases have not even been able to identify) to crack the key using conceivable computing power and resources. Figure 7.4 shows the graphic of an adversary attempting to break into either the cloud or a server or database and has to try all possible combinations of the keys to be able to even identify a combination that could successfully allow access to the system.

7.5.2 Cryptanalysis

Cryptanalysis is the opposite of cryptography and is the study of techniques to convert ciphertext into plaintext by identifying the key used for encryption or even matching possible plaintexts that could have been encrypted to that specific ciphertext, etc. There can be cases where the plaintext or key can be revealed by a skilled professional even without having the knowledge of the encryption technique that was used for generating the ciphertext.

There are multiple variants of this attack based on the outcome perceived (cryptographic keys or the message that was encoded). A small subset of such attacks is described below:

7.5.2.1 Known Plaintext Attack

In this attack scenario, the attacker/analyst has the ciphertext and multiple plaintext values. The analyst will have to compare the ciphertext and the plaintext and guess the key that could have been used for this operation and thus compromise the system as the key determined could be used to decipher other blocks of encrypted content.

7.5.2.2 Differential Cryptanalysis

This is a variant of the abovementioned plaintext attack where the analyst will continuously encrypt many different plaintext options that are available and compare the results with the cipher texts to reveal the value of the plaintext that was encrypted. Since this is a technique that relies on the difference between the multiple combinations, it thus gets the name as differential cryptanalysis.

These types of attacks have been identified to have great success against systems that are built in hardware. Since the hardware can go through phases of wear and tear, there are possibilities of errors or minor imperfections creeping in that could be observed to identify the key that was used internally and even reveal the plaintext in some cases.

7.5.2.3 Differential Power Analysis

This technique involves the careful observation of minor details including the timing or operation, power signatures, etc. and to determine the secret key that was possibly used in the encryption process.

7.5.3 Systems Based Attack

Finally, the other way where we are not concerned about the algorithm or the key but would like to impact the system involved in the cryptographic activity is to directly attack the system and make it incapable of being able to run/execute the cryptographic algorithm thus making the system unavailable. In this case, the algorithm is usually not altered or impacted.

7.6 Final Remarks

Symmetric key cryptography (or symmetric encryption) uses a single cryptographic key between the sender and receiver of an encrypted message. Such algorithms and systems are being used for decades and currently are used as a part of hybrid systems as they are very efficient and quick. Also, the added advantage of them being embedded into hardware modules has seen greater revolutionary applications.

References

1. What are block ciphers? WolfSSL. (2014, December 19). What is a block cipher? Retrieved April 07, 2021, from https://www.wolfssl.com/what-is-a-block-cipher/
2. Announcing the ADVANCED ENCRYPTION STANDARD (AES) - NIST. (2001, November 26). Retrieved June 10, 2020, from https://nvlpubs.nist.gov/nistpubs/FIPS/NIST.FIPS.197.pdf
3. DATA ENCRYPTION STANDARD (DES). (n.d.). Retrieved from https://csrc.nist.gov/csrc/media/publications/fips/46/3/archive/1999-10-25/documents/fips46-3.pdf
4. Initialization vector. (n.d.). Retrieved April 07, 2021, from https://cryptography.fandom.com/wiki/Initialization_vector

Chapter 8
Asymmetric Key Cryptography

The symmetric key cryptosystems have certain drawbacks which were highlighted in the previous chapter. To overcome the major disadvantage related to the compromise of one key being a potential threat to the whole system as it becomes a single point of failure, the Asymmetric key cryptographic techniques introduce two sets of keys: one that is publicly available and another that is a secret or private key that is retained with the communicating entities. It is considerable complex and expensive compared to symmetric key cryptography.

Each of the entities and specially the receiver of information generates a pair of keys and terms it as public key (pk) and private key (sk). The entity then broadcasts into the network the generated public key. So any entity that would like to send a message/communicate with this entity will use that public key to encrypt the message and route it to the receiver who will then have to use the private key to decrypt the message. The important aspect here is that a message that is encrypted with a public key can only be decrypted by the corresponding private key and not any other key. So any malicious entity that intercepts the communication and gains access of the encrypted message will be unable to decrypt and identify the message being transmitted.

Thus the only way to break the system will be to identify the private key. Another important advantage here is that the two entities will not have to previously agree on a key or even think about a secure mode of sending the keys that can be used for encryption or decryption as was the case in the symmetric key cryptosystems.

8.1 What Is Public Key Cryptosystem?

Asymmetric cryptography encrypts plain text messages using mathematical permutations as well; however, it encrypts and decrypts messages using two separate permutations, still recognized as keys. A public key that can be exchanged with anyone is used to encrypt messages in asymmetric cryptography, while a private key recognized only by the recipient is used to decrypt them.

Critically, computing the public key from the private key should be relatively simple, but generating the private key from the public key should be nearly impossible. RSA, ECC, and Diffie-Hellman are three common mathematical permutations that accomplish this today: algorithm is different, but they all obey the same basic principles. The RSA 2048 bit algorithm, for example, creates two 1024-bit prime numbers at random and multiplies them together. The public key is the result of that equation, while the private key is the result of the two prime numbers that generated it.

Since we are using the public key for the encryption and eliminating any need for key exchange, these systems are typically also interchangeable called as Public Key Infrastructure (PKI).

8.2 Advantages of Asymmetric Key Cryptosystems

Asymmetric key cryptography is usually considered an extension and enhancement over the symmetric key cryptosystems and hence the advantages of these algorithms are to alleviate the drawbacks of the symmetric key systems. Starting with the most important drawback of secure key transmission which is not required in the asymmetric key cryptosystems. Since there is no need for exchanging keys prior to communication, it eliminates the key distribution problem considerably.

Another important advantage of using asymmetric key cryptography is the increased security that can be achieved as the secret key/private key that will be used in the decryption process is not transmitted or made public knowledge and hence is away from the attackers reach. Thus simple passive attacks like eavesdropping leading to replay attacks or masquerading can be alleviated to some extent. One other more important advantage of the PKI system is the support provided for digital signatures and hence uphold non-repudiation.

8.3 Drawbacks of PKI

A critical drawback in comparison to the symmetric key cryptosystems is that the speed of operation and the cost of operation are both higher. To add to it, the introduction of multiple keys in the system increases the overall system complexity. Also, in some cases, the distribution of public keys can become complex.

There are other drawbacks in terms of the use of digital signatures and certificates that will have to be verified at the receiver's end and can be a complex process and involve considerable latency and network overhead apart from the cost involved in requesting and receiving certificates and maintaining certificate revocation lists (CRLs).

8.4 Concluding Remarks

In this chapter we briefly introduced the concept of asymmetric key cryptography and the advantages of this system over the symmetric cryptographic variant. There are various types of asymmetric key cryptosystems that have not been discussed as a part of this book.

In the next section, we will extend our knowledge of cryptosystems by introducing some of the modern encryption schemes and their benefits and disadvantages and how they motivated us toward the building of a tamper evident scheme.

Part IV
Modern Encryption Schemes

Chapter 9
Homomorphic Encryption

Statistical methods and computational capabilities have grown tremendously in the past few decades. Along with the growing interest in Artificial Intelligence and Machine Learning, there is also significant development in the field of Cloud Computing and Big data. Nowadays, we use a wide range of tools that have given us the power to stripe this data, process them and analyze or even predict outcomes using this data. Large volumes of data also have a downside in the amount of resources it utilizes in terms of storage and computation. This has led to the rise in the use of cloud services which provide a multitude of services that can benefit in transforming the raw data into useful information while also providing storage capabilities at a cost.

As more and more systems and service providers are using the cloud technologies to support their infrastructure and even provide recommendations, it is required to assess if the data that we share with these service providers is secure or can infiltrate into our privacy and leak information about us that can cause losses in various forms. There have already been many evidences of the devastating effects that even misconfiguration of cloud services can have in revealing personally identifiable information.

Homomorphic encryption was introduced in 1978 after RSA was presented by Rivest, Shamir, and Adleman. The homomorphic properties allow the manipulation of encrypted data using arithmetic operations while sharing the information with the guarantee that only the owner of the required private keys being able to decrypt the message and get hold of the contents. Figure 9.1 depicts a scenario wherein a user of cloud services can use encrypted versions of the request in order to receive an encrypted output that can be decrypted to get the original message.

Homomorphic encryption techniques provide a way to ascertain a limited level of privacy as well by providing a simple and efficient solution. The technique involves encryption of the data that is sent to the cloud. However, we have to ensure that we use appropriate encryption methods as simple symmetric encryption like the block ciphers obtained using AES encryption schemes cannot support manipulation

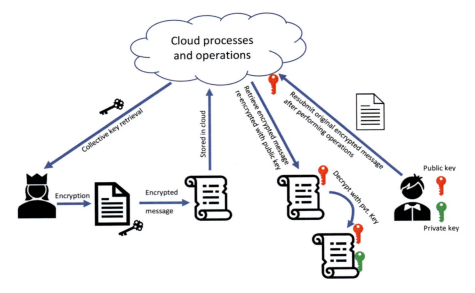

Fig. 9.1 Homomorphic encryption

without first decrypting the ciphertext. Public key encryption techniques using asymmetric keys for encryption and decryption provided a solution to this by ensuring that anyone with a public key can encrypt the message but only the rightful owner/recipient with the necessary private keys can decrypt the message.

9.1 Applications of Homomorphic Encryption

There are many applications for the properties that homomorphic encryption has to offer. A list of such applications is as discussed below.

9.1.1 Data Security in the Cloud

With the many useful resources and benefits that cloud based services offer, many organizations have been opting to use such services and free their local resources of some burden. However, in most conditions, the cloud is not a trustworthy environment. In such situations, the use of homomorphic encryption can aid in securing the data even with the added benefits of performing complex calculations. This scheme also enhances the search for ciphered information that can later be decrypted without any compromise to the integrity of the data at any authorized consumer side.

9.1 Applications of Homomorphic Encryption

Fig. 9.2 Data stored in the cloud is secured using homomorphic encryption

Homomorphic encryption in the cloud is in its nascency and there is still significant work to be done in this field for this technique to be adopted in a large scale by individuals and large corporations. There are some schemes of homomorphic encryption that are still not a viable option and are still in the state of research and are very expensive to even setup. However, a partial homomorphic encryption scheme (PHE) has been said to provide enhanced security in cloud systems and are currently being used by some cloud service providers. In due course of time, these schemes are expected to be extensively used not only in cloud computing but for other applications as well (Fig. 9.2).

Homomorphic encryption also has certain vulnerabilities which may jeopardize its future. FHE, for instance, can accommodate an outsized number of operations. However, considering the very fact that this is often also the very concept of FHE, an open problem is that FHE has drawbacks while supporting large number of operations/functions. The reason for this is because of the possibility that two operations can cancel one another, rendering the overall protection gain useless.

There are still some problems with homomorphic encryption which can impact its future. For instance, FHE can support quite one operation. However, an open issue is that FHE also has limitations on supporting a good range of operations/functions, albeit this is often the very definition of FHE. this is often because two operations might be wont to cancel one another out and make the safety pointless. Because a malicious user can just subtract 1 until the encrypted value is zero, giving the solution. This results in another issue, currently FHE is assumed of because the perfect solution; however, it must be considered on an application by application basis.

A one size fits all solution is not getting to be as secure as a scheme which is meant for the appliance in mind and eventually, by having homomorphic encryption protect user information/data, it stops cloud services from learning information about them. This will stop targeted ads, selling anonymous user data and lots of other ways cloud services make money albeit there is no cost to the top user. The difficulty is that albeit users want to be safer online; will they be willing to buy the cloud service, or will they like the free, unsecured service instead? These are just a few of the present issues that homomorphic encryption faces because it tries to become the longer term of security within the cloud.

9.1.2 Support to Data Analysis

The homomorphic encryption technique can be used in secure outsourcing of data even in the presence of untrusted channels. The integrity of the data is maintained until it is decrypted at the consumer site with appropriate credentials. This property is being utilized by many industries involved in the field of financial services, retail and ecommerce, information technology, and healthcare to allow people to use data without seeing its unencrypted values. As an example, homomorphic encryption has the capability to perform predictive analysis and hence provide services that can analyze. multiple types of data including medical data without the risk of exposing any personally identifiable information of the patient (Fig. 9.3).

Several studies that realize evaluating descriptive statistics using FHE are reported. Evaluating the quality descriptive statistics like the mean and variance from FHE ciphertexts is studied earlier. At the instant, many business associations are leveraging cloud services thanks to their specified simplicity and price effectiveness. There are, however, concerns over government inspection of all data due in security and attacks which they perceive that the service providers may breach by conniving assorted reminder insurgents. This is often seen as act of enervating the proper of the cloud providers to travel fully cloud computing. Despite this

Fig. 9.3 Data analysis in an untrusted environment is easier using homomorphic encryption

enervation, and since of huge accumulations of massive Data for analysis and with limited computational resources, most enterprises proffer to storing their data on the cloud. Doing so, nevertheless, has its adverse cost due to insufficient access controls are driving conversations about information security controls within the cloud. As an example, enterprise and Software as a Service (SaaS) providers have particularly high interest in using cryptographic techniques for shielding data within the cloud.

9.1.3 Enhancing Secure Ballot and Electoral Systems

Elections play a vital role in defining the future of a nation and the world. Many countries including the USA have opened for the citizens to cast their valuable votes electronically. Thus, there is a requirement for a system that can be transparent and secure while handling the votes cast effectively. Research has identified that homomorphic encryption techniques similar to that of Paillier encryption scheme can perform better in such circumstances because of the possibility to perform mathematical operations on the already encrypted votes and thus secure it from any kind of modifications and tampering. It also provides the advantage of being able to verify and validate the votes at any point in time.

The use of homomorphic encryption in the electronic voting use case is not just limited to encrypting the votes and hence securing them. These schemes can be used for the purpose of tallying and associating a vote to an individual. Since the vote is encrypted, there is no chance for an outsider without the appropriate privileges or authorization to be able to learn the contents of the vote. Hence an organization ascertained the task to identify the people who have voted and build the turnout statistics that would have to be published in the news or electoral websites. This can prevent duplicate votes and even possible fraud in terms of adversaries masquerading and casting the vote on behalf of some legitimate citizens (Fig. 9.4).

Practical variants of the homomorphic encryption scheme used in ballots also use zero-knowledge proofs (ZNP) wherein the voter can prove that the she/he voted and has cast a correct vote and encrypted it accordingly without actually having to

Fig. 9.4 A secure e-ballot

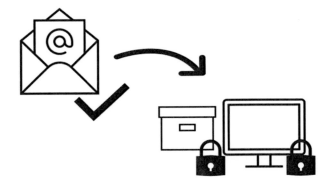

reveal what the value of the vote was. In addition, the validators can perform a set of operations that should give a certain value on completion and determine that the vote has been cast.

At the time of counting, with at least one honest validator, researchers have proposed the use of mixnets and onion routing concepts to mask the users and anonymize the users so that the votes are registered and counted without exposing the person who cast that specific vote. An open challenge though is to simultaneously find a way to notify the person who cast the vote that their vote has been successfully counted and considered as a part of the election results. There are still considerable advancements that are required in this application.

Overall, homomorphic encryption schemes have multiple advantages. However, the cost for implementation and operation has been a roadblock in their extensive adoption for various applications. However, research in this field has indicated a surge of interest in this direction and soon there could be more applications that extensively use these schemes.

9.2 Classification of Homomorphic Encryption: Examples

From the above use cases and discussion, we can conclude that homomorphic encryption has the potential to secure the integrity of information even in a chaotic environment with the addition that third-party users can manipulate the retrieved information which is already encrypted but will not be able to reveal or unveil the underlying content. Only the authorized user or data owner who possesses the private key for the initial encryption can decrypt the data and still view it without any tampering [1] (Fig. 9.5).

There are three major types/categories of homomorphic encryption with the difference between them being supported mathematical operations and the periodicity with which these operations can be performed/computed. The general flow of events in a system using homomorphic encryption is as shown in Fig. 9.6

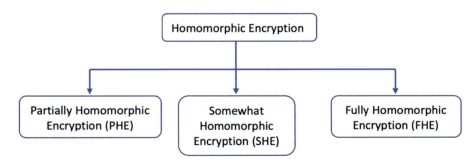

Fig. 9.5 Homomorphic encryption classification

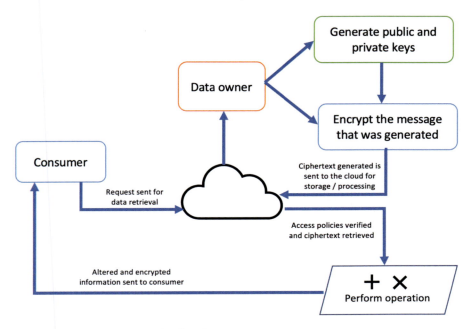

Fig. 9.6 Homomorphic encryption flowchart

9.2.1 Partially Homomorphic Encryption (PHE)

The first type of homomorphic encryption is called *Partially Homomorphic Encryption*. This type of homomorphic encryption allows for selected mathematical operations to be performed on the encrypted values that are retrieved from the source. The encryption schemes that fall under this category allow for multiplication to be performed on the ciphertext. RSA scheme which is traditionally used in securing connections by building SSL and TLS is a variant of PHE. Other important schemes that are categorized under PHE include ElGamal encryption which supports multiplication and Paillier encryption which uses addition operation [2].

9.2.2 Somewhat Homomorphic Encryption (SHE)

SHE is a scheme in which certain operations like addition or multiplication can be performed to a certain complexity. The complexity and the number of operations that can be performed are limited. SHE defines the initial stage of a fully homomorphic encryption scheme.

9.2.3 *Fully Homomorphic Encryption (FHE)*

Of the three categories of homomorphic encryption, the newest inclusion is the Fully homomorphic encryption (FHE). The initial idea of FHE was conceived by the American computer scientist Craig Gentry in 2009 [3]. Since its introduction, there has been multiple areas where potential applications for have been identified for this scheme. The benefit of using this approach is that it provides privacy primitives along with securing the information. As an enhancement of the previous two schemes, the FHE schemes can support both addition and multiplication as against just one of them by the schemes under SHE. FHE is said to possess capabilities to enhance and secure the current cloud computing systems and technology.

9.3 Problems of Importance in the Public Key Infrastructure

The adoption or even suggestion of a fully homomorphic encryption scheme took about three decades after its predecessor because of the practical difficulty in implementing such schemes which is true even to this date. Research thus far in this specific field has focused on theoretical optimizations and more often has a very strong assumption to work with. Exploration of FHE has made the implementations of SHE more robust over the past decade. However, there is considerable overhead in terms of both computational performance and size of parameters which has inhibited adoption for practical applications. However, large corporations are doing their part in creating libraries that can support the FHE and SHE schemes so that customers can use them for their applications.

9.4 Concluding Remarks

In this chapter we introduced the concepts of homomorphic encryption and discussed some of the applications where these schemes are used or even have a potential to be used. We also identified the various types of homomorphic encryption, their features, common types under each of them, and the differences among them. The following chapters discuss the various popularly known and used homomorphic encryption schemes and the implementation details of the same with the manner in which it is integrated in the design of TEDSP.

References

1. Crane, C. (2021, March 12). What is homomorphic encryption? Retrieved April 07, 2021, from https://www.thesslstore.com/blog/what-is-homomorphic-encryption/
2. What is homomorphic encryption? (2021, January 13). Retrieved April 07, 2021, from https://www.experfy.com/blog/bigdata-cloud/what-is-homomorphic-encryption/
3. Gentry, Craig. "Fully homomorphic encryption using ideal lattices." In Proceedings of the forty-first annual ACM symposium on Theory of computing, pp. 169–178. 2009.

Chapter 10
Popular Homomorphic Encryption Schemes

There are many applications that can benefit from the adoption of homomorphic encryption schemes. However, one of the front runners in terms of consideration for adoption is cloud computing systems. With the use of homomorphic encryption there have been use cases in this field that have been identified and tried that involve transmission of information to and from operators with no loss in integrity [1]. Figure 10.1 indicates the timeline of evolution of the homomorphic encryption schemes as described in [2] starting in 1976–2016. Researchers have been working on identifying ways to improve the adoption rate of these schemes and identify applications where its use is compelling and feasible in terms of the implementation and operational cost associated with it.

10.1 Goldwasser–Micali Method (GM Method)

In 1982, Shafi Goldwasser and Silvio Micali came up with the Goldwasser–Micali Method [3] which introduced a probabilistic approach to the traditional PKI system. This was introduced to eliminate the drawbacks caused by the deterministic nature of the current PKI schemes which made even brute force an effective adversarial tool to compromise the system. The GM method along with supporting homomorphism also makes it hard to compromise the system in a straightforward manner because of the introduction of randomness.

For a given plaintext m, a random string r is added to the and the receiver's public key and encryption is performed using this key as $E(m, r)$.

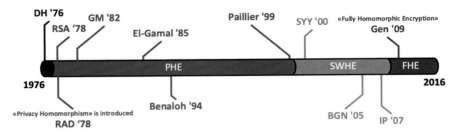

Fig. 10.1 Homomorphic encryption timeline. *Source*: This image is from - Acar, Abbas, Hidayet Aksu, A. Selcuk Uluagac, and Mauro Conti. "A survey on homomorphic encryption schemes: Theory and implementation." ACM Computing Surveys (CSUR) 51, no. 4 (2018): 1–35

At the receiver's end, the following steps are performed to complete the decryption process.

Algorithm 1: GM method: key generation

Result: Public key *pk* and Secret key *sk*
while *p and q are large random prime numbers such that* $p \neq q$ **do**
    Calculate $N = p * q$;
    pk_i =pseudosquare(p, q);
end
Identify u such that $u^2 \equiv a \pmod{N}$;
a is a value such that $\frac{a}{p} = \frac{a}{q} = -1$;
$pk = (N, a)$;
$sk = (p, q)$;

10.1.1 Encryption

Once the key generation phase is completed, the data producer to encrypt the data for a particular entity to be the recipient will have to do the steps mentioned in the algorithm below

---
Algorithm 2: GM method: encryption

Result: Ciphertext *c* for a specific consumer
Identify random *r*;
while *Message m is not null* **do**
    For each bit in *m*;
    if $m = 0$ **then**
        $c = r^2 \pmod{N}$;
    else
        $c = a * r^2 \pmod{N}$;
    end
end

10.1 Goldwasser–Micali Method (GM Method)

The value of r is chosen to be random such that adversaries in the system can be averted and prevent them from compromising the integrity of the message. The random values constitute all combinations of possible squares of modulo N. Once the ciphertext c is generated, it is routed to the consumer who will then decrypt it to reveal the message.

10.1.2 Decryption

Once a ciphertext value is received at the consumer end, the decryptor will use the private key components (sk that was previously computed) to identify the bit that was being transmitted. The decryptor thus computes the value of c/p and determines the following:

- $m = 0$ if $c/p = 1$
- $m = 1$ if $c/p = -1$

Overall, the GM method provided a new way of approaching and computing ciphertexts from plaintexts with enhanced security settings owing to the probabilistic nature. However, by today's standards it is not very efficient because of the computational complexity, but this method is used in the implementation of many homomorphic encryption schemes.

Figure 10.2 depicts the time taken for key generation, encryption, and decryption of messages having various sizes when using the Goldwasser–Micali method. The experiments show the performance of the scheme.

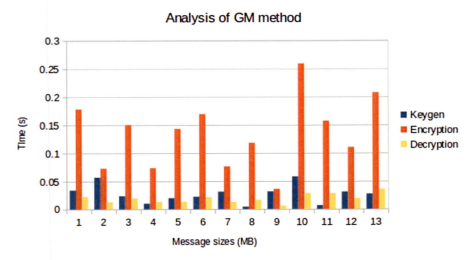

Fig. 10.2 Analysis of GM method

10.2 Paillier Encryption Scheme

Similar to the GM method, the Paillier scheme is a type of Partially Homomorphic Encryption. It was introduced in 1999 and named after Pascal Paillier [4]. It is built on crypto primitives that revolve around the key pairs generated. The keys are generated similar to the manner in which we do in a PKI based system. The uniqueness of the Paillier scheme is that as a PHE scheme, it supports "additive homomorphism," i.e., messages can be added together while they are encrypted and yet can be decrypted correctly without any loss of integrity.

To understand the working better let us consider A and B to be two communicating entities with A generating the data and transmitting it to B. Let us consider as a part of this process, A created two chunks of information labelled as m_1 and m_2. Let us also assume that the public key of the receiver B is known to be pk_B. The producer can encrypt m_1 and m_2 with pk_B such that $E(pk_B, m_1)$ giving $C(m_1)$ and $E(pk_B, m_2)$ giving $C(m_2)$. The producer can now add $C(m_1)$ and $C(m_2)$ and send it to B who will be able to correctly decrypt to individually obtain m_1 and m_2.

Figure 10.3 depicts the time taken for key generation, encryption, and decryption of messages having various sizes when using the Paillier method. The experiments show the performance of the scheme.

10.2.1 Benefits of the Additive Homomorphic Encryption

There are various advantages of homomorphism and in specific for additive homomorphism. If we take a specific example of secret electoral ballots, the use

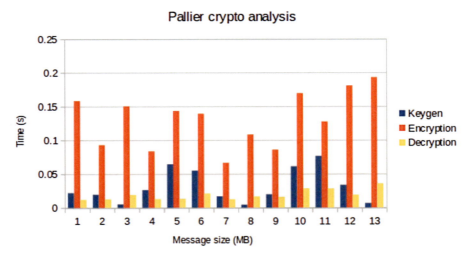

Fig. 10.3 Analysis of Paillier method

of additive homomorphism can add a considerable boost to maintaining integrity even while being able to accurately tally the votes that have been cast [5]. Additive homomorphism helps us achieve the basic requirement as we see for this case:

- The votes that have been cast should be encrypted not to reveal the actual person or party the vote has been cast for.
- The votes, however, should be counted and we would like to know if the person who voted was a legitimate citizen and not someone who is impersonating and committing a fraudulent transaction.
- Finally, there should be a way for the electoral team entrusted with counting to be able to identify the person voted for without being able to reveal whose vote is being checked.
- Also, if possible, even without the need for revealing the contents of each of the votes, if the board could determine the winner.

All the abovementioned requirements can be made possible by using additive homomorphism as:

- Voters receive a public key generated by the electoral board and use that to encrypt the votes they are casting.
- The encrypted votes are sent to the electoral commission and forwarded to the counting system/server that aggregates and validates the legitimacy of the votes. In this step, the system would categorize it into classes based on the encrypted value received.
- The electoral board now will be able to identify the sum of votes received by each candidate even without actually examining each of the votes in the ballot and thus be able to announce the winner of the elections.

Along with the abovementioned modifications, we could also employ the benefits of zero knowledge proof [6] through which the voters and even the counting section of the electoral board will be able to prove that a vote was cast and considered even without actually revealing the contents of the vote.

10.3 ElGamal Encryption Scheme

ElGamal encryption scheme is a variant of the public key cryptography. It was introduced in the year 1985 by Taher Elgamal [7]. It involves the use of probability to ensure that for a given message (plaintext), there can be various versions of ciphertext that can be derived. This crypto algorithm is based on cyclic groups and the security is dependent on the difficulty involved in breaking the logarithm in such a group. It is reliant on the fact that in a cyclic group given a^h and a^m, it is difficult to compute the value of a^{hm}.

The working on this cryptosystem on a high level can be defined as shown below for the various operations including key generation, encryption, and decryption.

Algorithm 3: ElGamal: key generation

Result: Receiver's public key *pk* and private key *sk*
Choose a large number q;
Compute the cyclic group F_q;
while F_q **do**
  Choose element g;
  Choose element a;
  $gcd(a, q) = 1$;
  Compute $h = g * a$
end
$pk = F, h = g^a, q$ and g;
$sk = a$;

Once the public and private keys are generated, the encryption of messages uses the public key that is received and is performed as highlighted in the below algorithm.

Algorithm 4: ElGamal: encryption

Result: Ciphertext c
while *Cyclic group F* **do**
  Choose element k such that $gcd(k, q) = 1$;
  Compute $p = g^k$;
  Compute $s = h^k = g^{ak}$ Multiply s with M;
end
$c = (p, M * s) = (gk, M * s)$;

For decryption,

- Calculate $s' = p * a = g^{ak}$
- Divide $M * s$ by s' to obtain the message M since in we can conclude that $s = s$

10.4 RSA Cryptosystem

RSA cryptosystem is one of the most commonly used encryption schemes today. The Internet security based on SSL/TLS uses RSA as the backbone. This cryptographic algorithm was introduced in 1978 by Ron Rivest, Adi Shamir, and Leonard Adleman on whose name the algorithm is famously called RSA [8].

It is a public key cryptosystem which is based on the difficulty of identifying factors of the multiplicative result of two large prime numbers. Compared to newer algorithms, RSA is an expensive in terms of time and space complexity but is still

10.5 Concluding Remarks

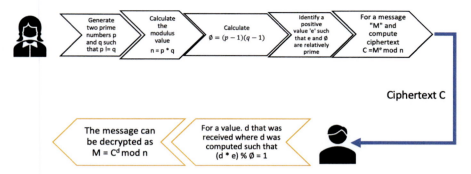

Fig. 10.4 RSA cryptosystem flowchart

a very common choice for key exchange, secure token generation and distribution, etc.

Figure 10.4 describes the steps involved in the key generation, encryption, and decryption of messages using the RSA algorithm.

In summary, the commonly used homomorphic encryption schemes and the mathematical operations they support are as mentioned in below table. Based on the requirement of the application for which these algorithms are to be employed, the best suited algorithm is selected from the list. The list mentioned in the table is not an extensive list but a small subset of the many schemes that exist with specific focus on PHE techniques.

Homomorphic scheme	Operation supported
Goldwasser–Micali	Addition
RSA	Multiplication
ElGamal	Multiplication
Paillier	Addition

10.5 Concluding Remarks

Homomorphic encryption has proved to be advantageous in many fields where there is a requirement for securing information. This chapter highlighted four famous homomorphic encryption schemes in RSA, Paillier, GM, and ElGamal. Each of these uses different mathematical operations and also applications in the real-world. In the next chapter, we shall see the industrial involvement in the design and use of homomorphic encryption schemes.

References

1. Ogburn, Monique, Claude Turner, and Pushkar Dahal. "Homomorphic encryption." Procedia Computer Science 20 (2013): 502–509
2. Pogiatzis, A. (2018, July 20). Homomorphic encryption - the holy grail of online confidentiality. Retrieved April 07, 2021, from https://medium.com/@apogiatzis/homomorphic-encryption-the-holy-grail-of-online-confidentiality-f6e505365039
3. Goldwasser, Shafi, and Silvio Micali. "Probabilistic encryption." Journal of computer and system sciences 28, no. 2 (1984): 270–299.
4. Paillier, Pascal. "Paillier Encryption and Signature Schemes." (2005): 453.
5. Paillier cryptosystem. (n.d.). Retrieved April 07, 2021, from https://paillier.daylightingsociety.org/about
6. Goldreich, Oded, and Yair Oren. "Definitions and properties of zero-knowledge proof systems." Journal of Cryptology 7, no. 1 (1994): 1–32.
7. ElGamal, Taher. "A public key cryptosystem and a signature scheme based on discrete logarithms." IEEE transactions on information theory 31, no. 4 (1985): 469–472.
8. RSA cybersecurity and Digital risk management solutions. (2021, March 29). Retrieved April 07, 2021, from https://www.rsa.com/en-us

Chapter 11
Industrial Involvement in the Use of Homomorphic Encryption

Homomorphic encryption as discussed in the previous chapters has been a topic of great interest in untrusted systems like the cloud systems. There are various other applications of using the system as well. However, when developers started incorporating the design details into a working prototype, they encountered the computational complexity that is involved in the successful deployment of the scheme.

Though the partially homomorphic and the somewhat homomorphic techniques have penetrated into being used in real-life applications, the fully homomorphic encryption technique has not been used for the large computational costs involved in its usage. It is also seen that the implementation of a fully functional homomorphic encryption scheme that can perform multiple operations and thus secure the system is also difficult.

After about a decade of being proposed, only recently have big organizations like IBM and Microsoft started developing the prototype that incorporates the design of a fully homomorphic encryption scheme. This chapter illustrates the various libraries that exist developed by these technology giants. A list of the open-source fully homomorphic encryption supporting libraries is as shown in below table.

Library name	Supporting organization	Language used
HElib	Developed by IBM (Open Source)	C++
SEAL	Developed by Microsoft (Open Source)	C++
TFHE	Open Source	C/C++
PALISADE	Duality Tech and collaborating Universities	C++
FHEW	Open Source	C++

11.1 IBM's HElib

HElib is a C/C++ based library for that implements homomorphic encryption specially focusing on the Brakerski–Gentry–Vaikuntanathan commonly known as the BGV scheme. The name HElib [1] itself is an acronym for Homomorphic Encryption library. This library was developed by IBM and is offered as an open-source library. Since its first version, there have been multiple optimizations that were put through to release the current version which is about 75 times faster than the first version that was released. They are, however, the first library to support FHE though still very slow for practical applications.

11.2 Microsoft SEAL

Microsoft Corporation developed its own library and has it available under an open-source license for individuals and organizations trying to implement or use homomorphic encryption schemes. With the SEAL (Simple Encrypted Arithmetic Library) [2], users will be able to perform computations on the encrypted data directly as is the inherent capability of homomorphic encryption schemes and test end-to-end use cases.

As an example, let us consider a common scenario involving cloud storage and retrieval. Before the use of homomorphic encryption schemes, in a traditional setup if data collected is sent to the cloud operator for any kind of processing and storage, the cloud services would have to decrypt the message and then process it before encrypting and then sending it back to the user when there is a request for retrieval. During the phase/duration when the data in cleartext is at risk of being compromised in-case an adversary manages breach the cloud security because of possible misconfigurations or other means. Even with policies like data not to be shared with third parties for monetary gains, there are cases where the third parties have had the opportunity to get access to data from cloud service providers. Figure 11.1 depicts this scenario where the user's information is vulnerable and at the risk of being compromised.

With the use of Microsoft's SEAL based Homomorphic encryption scheme, there is no need for the cloud service provider to decrypt the message and still be able to perform operations on the data in its encrypted form. This as discussed earlier is a major benefit of homomorphic encryption schemes and hence shields the data from the prying eyes of adversaries. Also, even if by some means information leaks to third parties who are not authorized to receive the information, the data retrieved by them would be in encrypted form and hence protects the integrity and confidentiality of the original message. Figure 11.2 depicts the scenario where homomorphic encryption is used with cloud storage and retrieval and hence provides enhanced security.

11.3 Other Homomorphic Encryption Library

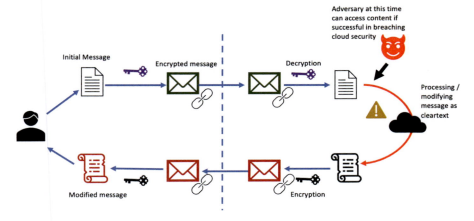

Fig. 11.1 Traditional cloud storage and computation

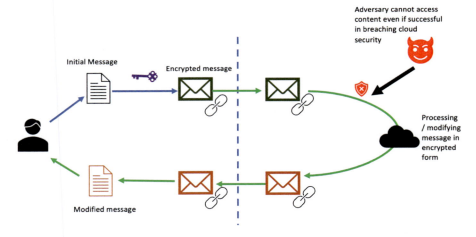

Fig. 11.2 Microsoft SEAL cloud storage and computation

11.3 Other Homomorphic Encryption Library

There are other organizations who either themselves or in collaboration with universities have implemented variants of homomorphic encryption libraries. The language used by most of these implementations is either C or C++ with some of them providing Python wrappers.

One such data security company that has developed a homomorphic encryption library is ENVEIL [3]. They use it as a part of their framework to protect the information in use.

Duality [4] is another company which in collaboration with multiple universities has created PALISADE a homomorphic encryption library in C++ with a Python wrapper. The company claims that with this tool, the users will be able to share data in raw form even with third parties without the fear of loss of integrity.

11.4 Final Remarks

Since 2016, there has been a great surge in industrial involvement in the design and release of fully homomorphic encryption support using libraries. These libraries are built in either C or C++ with wrappers to support other high-level languages and are being constantly modified to optimize their performance. This chapter discussed two such libraries in detail and other similar libraries which we believe will encourage the readers to try them and thus benefit from the various advantages that homomorphic encryption has to offer.

References

1. Helib: Helib documentation. (n.d.). Retrieved April 07, 2021, from http://homenc.github.io/HElib/
2. Microsoft seal: Fast and Easy-to-use homomorphic encryption library. (2020, November 12). Retrieved April 07, 2021, from https://www.microsoft.com/en-us/research/project/microsoft-seal/
3. Williams, E. (2020, April 21). A discussion on homomorphic encryption with Enveil CEO Ellison Anne Williams - Enveil: Encrypted Veil. Retrieved April 07, 2021, from https://www.enveil.com/news-press-posts//a-discussion-on-homomorphic-encryption-with-enveil-ceo-ellison-anne-williams
4. Building secure plus platform for secure collaboration on sensitive data. (2020, November 11). Retrieved April 07, 2021, from https://dualitytech.com/

Part V
Creating a Tamper Evident System for the Cyber Era

Chapter 12
Introduction

The important requirement in the current cyber age is the need for empowering security that will provide tamper evident information exchange, i.e., information can be exchanged among any entity without any hassles or fear of its integrity being compromised. This can to a certain extent be achieved by creating an ensemble-based approach that is built on the best practices and lessons learnt from these systems in designing solutions that can help enhance information integrity and confidentiality with minimum impact on system performance. The solutions thus designed will have to adhere to affordable time/space complexity and researchers are actively working toward achieving this.

In this chapter, we discuss the effects of combining various cryptographic solutions designed in the previous chapters with the benefits of public key cryptography and homomorphic encryption and concepts of non-malleable encryption under the hood of a switchable tamper evident system. Along with leveraging the benefits of existing techniques, the proposed scheme transitions them to newer levels to gain an advantage over the attackers. Simulation results discussed in the upcoming sections of this book prove the superior performance of the proposed scheme in providing cryptographic security.

12.1 Limitations of Existing Public Key Encryption Techniques

Public key cryptosystems are currently the most widely used cryptographic tools in many applications. However, the security properties and primitives that these provide are usually highly prone to attacks as has been seen over the past few decades. Thus we needed a paradigm shift and a transition that can incorporate the best practices from multiple approaches to provide the world with a new way of securing the confidentiality and integrity along with provenance of content irrespective of the storage location or the medium used for communication.

Some of the major limitations and disadvantages of the existing cryptographic techniques are as mentioned below.

12.1.1 Major Limitations

The major limitations of the existing state-of-art public key encryption techniques are as follows:

1. Failure to provide all-in-one solution with reduced trade-off between security and other performance factors.
2. Not being able to provide unconditional information safeguard by utilizing the existing computationally bounded trapdoor one-way functions.
3. The inability of efficient ciphertext size techniques using security parameter independent plaintext to resist active attacks due to the existence of malleability property.
4. Surprisingly, almost all homomorphic public key techniques are not supporting tamper evident security paradigm. The currently existing techniques might solve security issues, but they still suffer from ciphertext expansion problem or overall running time.

12.1.2 Can a Shift from Deterministic to Probabilistic Approaches Provide a Better Outcome?

Despite of several prominent advantages, the state-of-art techniques in trapdoor one-way function-based public key cryptography suffer from the following problems:

1. No computation dependent techniques assure perfect secrecy (or unconditional secrecy).
2. Very less focus on security independent plaintext techniques (except Goldwasser–Micali family of constructions).
3. The injective trapdoor functions are widely used in public key cryptosystems, whereas the lossy trapdoor functions are widely used in one-way function constructions. No composite surjective, composite injective, or combinations of them can be incorporated to achieve an efficient public key cryptosystem.
4. Very few tamper evident support public key cryptosystems are available for secure information storage and retrieval applications.

Therefore, this multifold limitation has seriously undermined the construction of all-in-one security solution by many promising encryption techniques.

12.2 Motivation: Our Major Proposition

In order to overcome the secret sharing over insecure channel of symmetric encryption, the concept of asymmetric encryption was proposed. We would need the asymmetric schemes to support the following features:

- Be able to retrieve ciphertexts that have been expanded
- Operate on the ciphertexts without having knowledge of the actual plaintext and still having a say in the encryption scenarios
- All the inherent security properties and features and prevention of typical cryptanalysis

12.3 Tamper Evident Solutions

A novel approach to using a combination of the benefits that uses PKI and homomorphic encryption with concepts of non-malleable encryption is proposed with the initial experiments showing an enhanced security properties. On a high level, the proposed approach is similar to an encryption blackbox constituting a switch (that is based on a psuedorandom generator and dependent on certain parameters) that chooses one of the abovementioned encryption schemes to compute the ciphertext of a given message.

TEDSP has the unique feature of switching between multiple encryption schemes by carefully selecting the appropriate combination of public key as shown in Fig. 12.1. This switching between the two encryption techniques has many benefits as listed below:

1. The benefits and properties of both the encryption schemes can be harnessed to get a robust system.
2. Free plaintext space independent of the security parameter.

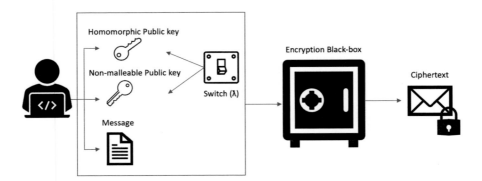

Fig. 12.1 Proposed switchably malleable encryption approach of TED-SP

3. The switch can assure the probabilistic property and thereby secure the data from being compromised by attackers
4. The use of both the mechanisms interchangeably compensates for the time complexity, thus making it a feasible option. Other optimization techniques are required.
5. TED-SP can support information exchanges securely even if it is between entities without prior communications.
6. Inherent tamper evidence to the underlying secret information.

12.3.1 Working of the Proposed Scheme

If P_{KH}, P_{KP}, and P_{KN} are the homomorphic, public key, and non-malleable public key components, respectively, the important aspect of this design is the random generation of the specific property supported public key components and the appropriate function selection during encryption process. For each encryption instance, these property specific public key components are randomly generated when a random function generator *Switch* is invoked, which generates a value N and a switching flag λ which is either 0 or 1 or 2 with 0 indicating homomorphic, 1 indicating Public key crypto-algorithm, and 2 indicating non-malleability and outputs the ciphertext computed accordingly as shown in Fig. 12.1.

12.3.2 Advantages of TEDSP

The major benefit of using TED-SP is that it can be applied without any changes to the existing setup and hence is easily scalable without the need for any kind of communication between the entities or even works well in a highly chaotic environment. This scheme is also built on the best practices of the abovementioned techniques and hence overcomes the individual disadvantages of the systems. Though the time taken for homomorphic cryptographic schemes is high, because of the use of faster algorithms in conjunction, the overall time taken by the entire system over multiple iterations is comparable to the existing schemes with the added advantage of better security properties.

12.4 Impact on Information Storage and Retrieval Applications

The proposed model provides a tamper evidence to the stored data when it is used to store the information securely over the untrusted interfaces and environments.

There are a couple of use cases that have to be described here and will be updated soon.

12.5 Final Remarks

In this chapter we discuss the underlying concepts of the proposed tamper evident security protocol. The TEDSP model combines the benefits of the deterministic and probabilistic approaches of using the homomorphic encryption and non-malleable encryption schemes. This provides a highly robust system that can be trusted by the users even in highly untrusted environments like the cloud services.

The chapter also provides an account of the technology transition that is required and can be achieved using the proposed TEDSP protocol. The system can scale seamlessly with the addition of more homomorphic encryption techniques that can perform even higher complex computations in order to secure the underlying information.

Part VI
Feasibility and Performance Analysis

Chapter 13
Implementation Details

In order to see the effectiveness of the proposed idea in the previous chapter, we are implementing a Python-based library called *PyTEDSP*. Experiments were run on the implemented library to identify the performance of TED-SP and compare it with other existing works. The preliminary results before and after optimization are shown in the graphs under the Simulation Results section. As discussed in the design of TED-SP, the experimentation environment incorporates the fully homomorphic and non-malleable encryption schemes and the functional parameterized switch.

13.1 Details of the Code Implemented

The main advantage of TEDSP is the use of the parameterized switch that switches between the use of homomorphic or non-malleable cryptographic methods for the encryption. The random switching leads to a probabilistic approach wherein the attacker who is looking at the blackbox of the system is unsure of the algorithm used for the encryption. With the availability of multiple well-known techniques of homomorphic encryption, the switch does better in randomizing the homomorphic encryption type used as well if the homomorphic encryption techniques are the outcome of the randomized switch.

This in turn increases the entropy in the system and decreases the probability of an attacker even with infinite computational capability to be unable to break the scheme and thus retrieve the message. Since we are using a combination of probabilistic and deterministic homomorphic encryption schemes along with the non-malleable techniques, the computation time in the form of key generation, encryption, and decryption is comparable to that of existing schemes. TEDSP in the worst case is as complex as the underlying algorithms are while providing a highly robust system that is *Tamper Evident*.

The code snippet of the parameterized switch is as follows:

Algorithm 5: Parameterized switch design

Result: Switch output
while *Range* **do**
    Define randomnumbergenerator in the range;
    while λ **do**
        Generate the switching frequency;
        Identify the Switch parameters;
    end
end
Switch outcomes based on parameters;

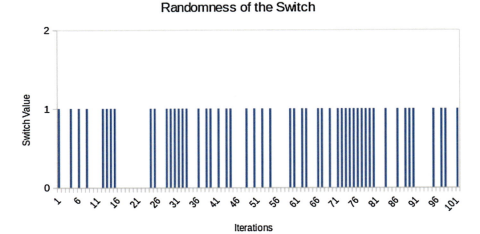

Fig. 13.1 Comparison of decryption times in low resource platform

The probability of picking a specific parameter that inclines to homomorphic or malleable cryptographic primitives is highly random and makes this proposed cryptosystem highly robust. The switching operation being highly randomized, it is difficult for any adversary to be able to predict if the ciphertext has been produced using the homomorphic encryption scheme or the non-malleable encryption scheme (Fig. 13.1).

The homomorphic encryption code involves the use of three famous homomorphic encryption schemes, namely RSA, Paillier, and the GM method. Implicitly, if the homomorphic encryption scheme is selected, in order to increase the randomness of the operation and hence make it even more difficult for the adversary trying to attack the system, we have the switch function again to select one of the three algorithms. Below is the high-level code design option for this proposed design and is just an abstract code snippet. (Note: the code provided below is not complete and will not compile.)

Algorithm 6: TEDSP design

Result: Ciphertext c
Define Switch based on parameters **while** *Switch* **do**
    if $\lambda = 0$ **then**
        if $t = 0$ **then**
            | Execute RSA algorithm;
        else
            | Execute ElGamal algorithm;
        end
    else
        if $m = 0$ **then**
            | Execute GM algorithm;
        else
            | Execute Paillier algorithm;
        end
    end
end
c = Generated Ciphertext;
Send c, nonce and uniqueid;

13.2 Comparison of Performance

Since this is the first attempt to combine the benefits of homomorphism and non-malleability using the TEDSP, we do not have an existing library that we can perform our experiments over. However, as a proxy, we use the implemented code in trying to analyze the performance in terms of common metrics to measure the key generation, encryption, and decryption. The results show the performance with optimized and unoptimized code in the library. The experiments were also run in a low resource platform like a Raspberry Pi 3B.

Figure 13.2 shows the time taken for encryption operation when the experiments were run using the non-optimized versions of the implemented library for varying bit sizes. The homomorphic operations included in these runs were addition, multiplication, and a combination of the operations while comparing the proposed scheme with the existing schemes. The key sizes used vary from 1K to 4K. It can be seen that the combination of addition and multiplication when performed on the ciphertexts increases the encryption time. However, it makes the system less vulnerable to being compromised.

Figure 13.3 showcases the results of running the system under the constraint of using just addition and multiplication operations in the native libraries to that

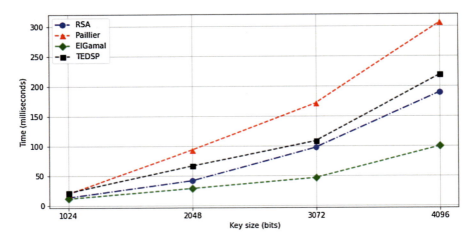

Fig. 13.2 Comparison of encryption times without optimization

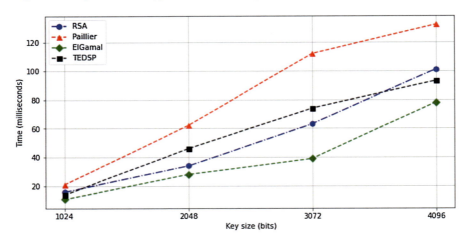

Fig. 13.3 Comparison of encryption times after optimization

of TEDSP. It can be seen that the time taken for the encryption using TEDSP is comparable to the exiting algorithms based on the switching operation. However, using other optimizations can improve the performance of the system overall with the added advantage of not being easy to break the cryptosystem.

Figure 13.4 shows the performance of the algorithm in comparison with the other algorithms in the electronic voting test case simulation. The graph depicts the time taken by each of the algorithms when the number of voters is varied. The performance of the proposed scheme is comparable to the other schemes while giving better security and being harder to break.

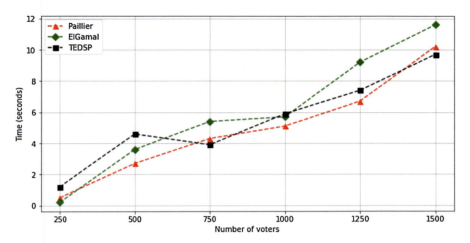

Fig. 13.4 Comparison of encryption times in low resource platform

13.3 Evaluation

The experiments run on the implemented library has also shown greater resistance to common security attacks on the traditional implementation of the individual PKI based algorithms. The results confirm that the TEDSP approach that combines the benefits of the deterministic and probabilistic approaches of homomorphic encryption is successful in delivering better results that are highly fault-tolerant. The system also makes it computationally hard for an adversary to break into the system.

It should be noted that the worst case performance in terms of time taken for the cryptographic operations like encryption or decryption of the plaintext message is equivalent to that of the time taken by using the traditional RSA approach. The biggest gain is the achievement in terms of being unable to break into the system

13.4 Final Remarks

The addition of a randomly operating probabilistic switch that chooses the use of either homomorphic or non-malleable cryptographic techniques ensures that the system is highly robust and as the name suggests "tamper evident" and performs very well even in the most untrusted environments.

There can be many combinations of these algorithms that can be used in effectively securing the contents in cloud storage, on the go edge processing servers, localized systems and even in places with profound adversarial activity. The results depicted in this chapter highlight the benefits of using this system and the increased difficulty as an adversary to be able to compromise such a system and be able to maliciously perform activities on it.

Part VII
Future Directions

Chapter 14
Quantum Cryptography

Contributed by Dr. Mario Mastriani
The content mentioned below in this chapter is contributed by Dr. Mario Mastriani based on the work conducted by him over the past decade. The authors of this book would like to thank Dr. Mastriani for his contribution and permissions to add this work in the book. As confirmed by Dr. Mastriani, the ideas specified below have not been published anywhere but are only discussed internally and stored in repositories like Academia.edu (https://www.academia.edu/39217563/Is_instantaneous_quantum_Internet_possible), and as the author and owner, he has the privilege to publish the content.

We would like to thank the author for his contribution and permission to use this work as a part of this book.

This is a summary of the work by Prof. Mario Mastriani over the past decade and is acknowledged in the references thanking the co-authors of the papers.

This chapter is inspired by the work performed by the co-author SS Iyengar along with his colleagues who have been focusing on various aspects of quantum communication. In this chapter we have added a novel futuristic approach to realizing cryptography using quantum computing techniques. The authors though have been a pioneer in this area and have published many articles in conferences, journals, and even books in related topics. We would like to thank them for their work and allowing us to include it as a part of this book.

This chapter summarizes computing in a quantum structure for a variety of problems and more specifically the following computational applications Quantum Cryptography [1–3] is one of the most important tools inside Quantum Communications [4–8] toolbox, where all techniques used for data communication are based on principles of Quantum Mechanics [9, 10]. Quantum Key Distribution (QKD) [11, 12] is the maxi-mum exponent of Quantum Cryptography. Specifically, QKD is responsible for the creation and distribution of a public key between a sender and a receiver, through which both parties can use it to encrypt and decrypt the message to be transmitted, respectively.

Next, the most modern Quantum Cryptography technique will be exposed to evaluate its projection over time and thus establish the most probable trends in terms of data security applied to IoT, Blockchain, and so on.

14.1 Prolegomenous on Quantum Key Distribution (QKD)

There are fundamentally two kinds of QKD protocols:

- **Preparation and measurement protocols:** also known as photon polarized states protocols, they fundamentally emphasize the collapse of the wave function after a quantum measurement, since in Quantum Mechanics [9] the mere act of measuring alters the state of what is measured. Then, this simple fact can be used to detect the presence of a hacker in the information transmission channel, since the hacker must necessarily measure if he wants to intercept the channel communication. This alteration due to measurement can also be used to calculate the amount of information that has been intercepted. Furthermore, these protocols generally have a classic channel between sender and receiver to verify the integrity of the key.
- **Entangled photon pairs protocols:** quantum entanglement [13–15] is a phenomenon whereby a single wave function represents one (or more) entangled particles, which lose their state of individual representation by their own wave functions, at least for the duration of the entanglement. This curious phenomenon apparently violates the local realism and causality [16, 17], since it is intrinsically monogamous [13]. However, if we make a local measurement of the state (spin orientation) of one of the two entangled particles, there is an instant notification to its counterpart, even when both are at a distance where said notification required a trip at a faster-than-light speed [18–23] to be instantaneous. When a measurement is made on one of the entangled particles, the collective wave function collapses and each particle becomes independent with their respective individual wave function. Therefore, any collapse of the collective wave function will imply that a measurement has been made. If this can be verified through a classic channel that has not been made by the sender or receiver, then it has to be a hacker.

Using the second family of protocols instead of the first one has the following advantages, which are literally extracted from [24]:

- Empty pulses can be removed, since if one entangled photon is detected, then its counterpart (i.e., the other entangled photon) must be present. An automatic consequence of this, at least at a first glance, is to have a 100% certainty of emitting a non-empty pulse. This is beneficial only due to the fact that currently available single photon detectors have a high dark count probability, although the difficulty of always collecting both photons of the same pair somewhat reduces this advantage.

- Photon pairs have additional advantages to that of avoiding multi-photon pulses, since the probability that a non-empty pulse contains more than one photon is essentially the same for weak pulses and for photon pairs, for a given mean photon number. However, the fact that passive state preparation can be implemented prevents multi-photon splitting attacks [25, 26], considering that the different photon pairs are independent. Then, even if multiple photon pairs do not increase the hacker's information, they lead to an increase in the Quantum Bit Error Rate (QBER) [25].
- Using entangled photons pairs prevents unintended information leakage in unused degrees of freedom [27]. Observing a QBER smaller than approximately 15
- Entangled photons offer interesting possibilities in the context of cryptographic optical networks. The photon pair source can indeed be operated by a key provider and situated somewhere in between potential quantum cryptography customers. In this case, the operator of the source has no way to get any information about the key obtained by sender and receiver.

In a nutshell [28], the main advantages of entanglement based QKD are the conceptual beauty, the immunity against photon number splitting attacks, and the potential lack of empty pulses. Additionally, they offer interesting possibilities in the context of cryptographic optical networks, since a key provider with a few sources of entangled photons can operate such a network with high redundancy, while the disadvantage clearly is the increased experimental difficulty. Currently, the advantages are in favor of the version with entanglement, in particular, for an interesting and practical application for the security of all types of information systems, which will be seen in a later subsection and which is known as the Virtual Entanglement procedure [29, 30].

On the other hand, the two kinds of protocols mentioned above can each be further divided into three families of protocols [28]:

- discrete variable
- continuous variable
- distributed phase reference coding

Discrete variable protocols were the first to be created, and they remain to be currently the most widely implemented in practice. The other two families are mainly concerned with over-coming practical limitations of experiments. Finally, two protocols, that use discrete variable coding, will be described in Sects. 14.3 and 14.4.

14.2 A Primer on Quantum Information Processing

For pure states, that is, states on the Bloch's sphere [31–33] of Figure 24.1, any wave function

$$|\psi\rangle = \alpha|0\rangle + \beta|1\rangle \tag{14.1}$$

arises from the superposition of the so-called Computational Basis States (CBS) and qubit basis states $\{|0\rangle, |1\rangle\}$, which are located at the poles of the already mentioned sphere with $|\alpha|^2 + |\beta|^2 = 1$, such that $\alpha \wedge \beta \in \mathbb{C}$ of a Hilbert's space [31]. Strictly, the complete wave function will be

$$|\psi\rangle = e^{i\gamma}\left(\cos\frac{\theta}{2}|0\rangle + e^{i\phi}\sin\frac{\theta}{2}|1\rangle\right) = e^{i\gamma}\left(\cos\frac{\theta}{2}|0\rangle + (\cos\phi + i\sin\phi)\sin\frac{\theta}{2}|1\rangle\right) \quad (14.2)$$

where $0 \leq \theta \leq \pi, 0 \leq \phi \leq 2\pi$ [31]. However, we can ignore the factor $e^{i\gamma}$ of Eq. (14.2), because it has no observable effects [40], and for that reason we can effectively write

$$|\psi\rangle = \cos\frac{\theta}{2}|0\rangle + e^{i\phi}\sin\frac{\theta}{2}|1\rangle \quad (14.3)$$

with $\alpha = \cos\frac{\theta}{2}$ and $\beta = e^{i\phi}\sin\frac{\theta}{2}$, being then equal to the Eq. (14.2). The numbers θ and ϕ define a point on the unit three-dimensional sphere, as shown in Fig. 14.1.

Now, we can represent the mentioned poles in different ways. In fact, we introduce an alternative and very useful version for the task to be developed, which is a scalar way to represent the poles based on the orientation of the spins:

$$Spin\,up = |0\rangle = \begin{bmatrix} 1 \\ 0 \end{bmatrix} = North\,Pole \quad (14.4)$$

$$Spin\,down = |1\rangle = \begin{bmatrix} 0 \\ 1 \end{bmatrix} = South\,Pole \quad (14.5)$$

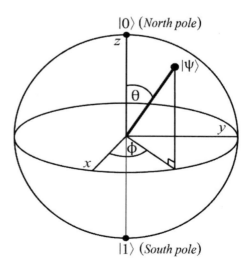

Fig. 14.1 Bloch's sphere

14.2 A Primer on Quantum Information Processing

We can see both spins (up and down) in the poles of Fig. 14.1. Moreover, and based on CBS, we can define another basis, which will be very useful for the rest of this chapter as well as for quantum information in general [31], and quantum teleportation and superdense coding in particular [6, 8, 13, 31, 34–38]. Taking into account the interaction of two subsystems A and B, considering that their components are pure states and using their scalar versions here too, we will obtain,

$$|0^A\rangle \otimes |0^B\rangle = |0^A\rangle|0^B\rangle = |0^A, 0^B\rangle = |0^A 0^B\rangle \tag{14.6}$$

$$|1^A\rangle \otimes |1^B\rangle = |1^A\rangle|1^B\rangle = |1^A, 1^B\rangle = |1^A 1^B\rangle \tag{14.7}$$

where $\otimes$ is the Kronecker's product [31] where

$$|\psi_1\rangle \otimes |\psi_2\rangle = \begin{bmatrix}\alpha_1\\\beta_1\end{bmatrix} \otimes \begin{bmatrix}\alpha_2\\\beta_2\end{bmatrix} = \begin{bmatrix}\alpha_1\begin{bmatrix}\alpha_2\\\beta_2\end{bmatrix}\\\beta_1\begin{bmatrix}\alpha_2\\\beta_2\end{bmatrix}\end{bmatrix} = \begin{bmatrix}\alpha_1\alpha_2\\\alpha_1\beta_2\\\beta_1\alpha_2\\\beta_1\beta_2\end{bmatrix} \tag{14.8}$$

Consequently with this, and being

$$|00\rangle = \begin{bmatrix}1\\0\end{bmatrix} \otimes \begin{bmatrix}1\\0\end{bmatrix} = \begin{bmatrix}1\\0\\0\\0\end{bmatrix}, |11\rangle = \begin{bmatrix}0\\1\end{bmatrix} \otimes \begin{bmatrix}0\\1\end{bmatrix} = \begin{bmatrix}0\\0\\0\\1\end{bmatrix},$$

$$|01\rangle = \begin{bmatrix}1\\0\end{bmatrix} \otimes \begin{bmatrix}0\\1\end{bmatrix} = \begin{bmatrix}0\\0\\1\\0\end{bmatrix}, |10\rangle = \begin{bmatrix}0\\1\end{bmatrix} \otimes \begin{bmatrix}1\\0\end{bmatrix} = \begin{bmatrix}0\\1\\0\\0\end{bmatrix} \tag{14.9}$$

we are going to use them to build the famous Bell's bases [31–33], with 2-qubit vectors, the combined Hilbert space will be $H^{A \cup B} = H_2^A \otimes H_2^B$, and then we will have the following four vectors,

$$|\beta_{00}\rangle = |\Phi^+\rangle = \frac{1}{\sqrt{2}}(|00\rangle + |11\rangle) \tag{14.10}$$

$$|\beta_{10}\rangle = |\Phi^-\rangle = \frac{1}{\sqrt{2}}(|00\rangle - |11\rangle) \tag{14.11}$$

$$|\beta_{01}\rangle = |\Psi^+\rangle = \frac{1}{\sqrt{2}}(|01\rangle + |10\rangle), \tag{14.12}$$

$$|\beta_{11}\rangle = |\Psi^+\rangle = \frac{1}{\sqrt{2}}(|01\rangle + |10\rangle) \tag{14.13}$$

Formally, bases of Eq. (9) are obtained from Eqs. (4) and (5) and a pair of gates, i.e., Hadamard (H) and CNOT gates:

$$H = \frac{1}{\sqrt{2}} \begin{bmatrix} 1 & 1 \\ 1 & -1 \end{bmatrix} \tag{14.14}$$

such that,

$$|+\rangle = H|0\rangle = \frac{1}{\sqrt{2}} \begin{bmatrix} 1 & 1 \\ 1 & -1 \end{bmatrix} \begin{bmatrix} 1 \\ 0 \end{bmatrix} = \begin{bmatrix} \frac{\sqrt{2}}{2} \\ \frac{\sqrt{2}}{2} \end{bmatrix} \tag{14.15}$$

$$|-\rangle = H|1\rangle = \frac{1}{\sqrt{2}} \begin{bmatrix} 1 & 1 \\ 1 & -1 \end{bmatrix} \begin{bmatrix} 0 \\ 1 \end{bmatrix} = \begin{bmatrix} \frac{\sqrt{2}}{2} \\ \frac{-\sqrt{2}}{2} \end{bmatrix} \tag{14.16}$$

and

$$CNOT = \begin{bmatrix} 1 & 0 & 0 & 0 \\ 0 & 1 & 0 & 0 \\ 0 & 0 & 0 & 1 \\ 0 & 0 & 1 & 0 \end{bmatrix} \tag{14.17}$$

such that,

$$|\beta_{00}\rangle = CNOT(|+\rangle \otimes |0\rangle) \tag{14.18}$$

$$|\beta_{01}\rangle = CNOT(|+\rangle \otimes |1\rangle) \tag{14.19}$$

$$|\beta_{10}\rangle = CNOT(|-\rangle \otimes |0\rangle) \tag{14.20}$$

$$|\beta_{11}\rangle = CNOT(|-\rangle \otimes |1\rangle) \tag{14.21}$$

The Bell's bases represented by Eqs. (14.18)–(14.21) will be of fundamental importance in the rest of this chapter, in particular, will be used in Quantum Teleportation [34] and its derivatives.

14.3 QKD Based on Polarized Single Photon

The first protocol of this type was the famous BB84 due to Bennett and Gilles [39], which mainly uses photon polarization states to transmit the information. Two pairs of conjugate states are used in this protocol $\{(|0\rangle, |1\rangle); (|+\rangle, |-\rangle)\}$, i.e., phase encoded states, which are used by all the implementations of BB84. The sender (commonly called Alice) and the receiver (known as Bob) constitute both

14.3 QKD Based on Polarized Single Photon

ends of a quantum communication channel which allows the transmission of the quantum states. In the case of photons the quantum channel is constituted by an optical fiber or a link with an optical cannon in the sender and a telescope in the receiver with free space between them. In the case of optical fiber is necessary to use quantum repeaters each 50 km due to attenuation and loss of optical material. Additionally, a public classical channel is necessary (radio, phone, Internet, and so on) to authenticate the shared key. In fact, the quantum channel cannot be interfered with by a third party (i.e., a hacker); however, and obviously, the public classical channel needs to be authenticated [40, 41].

All the protocol's security relies on an encoding of information in non-orthogonal states. No-Cloning Theorem [42] establishes that these states cannot, in general, be measured without disturbing the original state [43, 44], giving rise to quantum indeterminacy.

In BB84, $|0\rangle$ is orthogonal to $|1\rangle$, and $|+\rangle$ is orthogonal to $|-\rangle$ [39–41], while both pairs, i.e., $(|0\rangle, |1\rangle)$ and $(|+\rangle, |-\rangle)$ are mutually conjugates between them. Therefore, this protocol can use both basis: $(|0\rangle, |1\rangle)$ and $(|+\rangle, |-\rangle)$, indistinctly.

The protocol begins with a quantum transmission, i.e., the sender creates a random bit (0 or 1) selecting randomly one of the two mentioned basis to transmit that bit. A photon polarization state is prepared depending both on the bit value and basis, see Fig. 14.2, where as an example the two first columns of this table mean that a 0 is encoded in the rectilinear basis (+) as a vertical polarization state (↑), and a 1 is encoded in the diagonal basis (x) as a 135° state (↗). Then, Alice transmits a single photon via the quantum channel in the state specified to Bob. This process is repeated from the random bit stage, where the sender must register the state, basis, and time of each photon sent as a natural part of this protocol.

According to the quantum indeterminacy that we mentioned above, no possible measurement distinguishes between four different polarization states if they are

sender's random bits	0	1	0	1	1	0	1	0
sender's random basis	+	+	+	×	×	×	+	×
photon polarization sent by sender	↑	→	↑	↘	↘	↗	→	↗
receiver's random measuring basis	+	×	×	×	+	×	+	+
photon polarization measured by receiver	↑	↗	↗	↘	→	↗	→	→
Public discussion on basis								
shared secret key	0			1		0	1	

Fig. 14.2 Preparation of a photon polarization state depending on both bit value and basis

not orthogonal each other, i.e., only it is possible a measurement with recovery of projections inside an orthogonal basis. As an example, if a measurement is made at the rectilinear basis, this fact will give a result horizontal or vertical. In fact, if the photon was created with a horizontal or vertical polarization (as a rectilinear eigenstate), then this measures the correct state, but if the photon was created with a polarization of 45° or 135° (diagonal eigenstates), then the rectilinear measurement instead returns either a horizontal or vertical polarization at random. Moreover, after this measurement, the photon is polarized in the state it was measured in (horizontal or vertical), losing in this way all the information about its initial polarization.

Since the receiver does not know the base on which the photons were encoded, all he can do is select a random base to measure, either rectilinear or diagonal. The receiver must do this for each photon he receives, recording the time, measurement basis used, and measurement result. Once the receiver has measured all the photons, he communicates with the sender through the classic public channel. The sender transmits the base on which each photon was sent, while the receiver transmits the base on which each was measured. Both discard photon (bit) measurements where the receiver used a different base, which is half on average, leaving half the bits as a shared key.

To verify the presence of a hacker, the sender and the receiver now compare a default subset of their remaining bit strings. If the hacker has obtained information on the polarization of the photons, this introduces errors in the receiver measurements, which are evidenced in the aforementioned comparison. Similar errors can be caused by other environmental conditions. In the case that a greater number of bits were to differ, both sender and receive interrupt the transfer of the key and try again, maybe with a different quantum channel, since the security of the key cannot be guaranteed.

14.4 QKD Based on Entangled Photon Pairs

The first protocol of quantum cryptography based on the creation and distribution of an entangled photon pair was proposed by Ekert in [45], E91. This protocol uses the Bell's inequalities [X] as an alarm, which indicates the presence of a hacker in the channel if the statistics associated with the Bell's Theorem is locally altered by a third party.

The sender creates an entangled photon pair, conserves one of the photons, and sends another one to the receiver via an optical fiber (which requires a quantum repeater each 50 km) or thanks to an optical cannon in the sender and a telescope in the receiver with free space between both [46].

This protocol fully exploits two properties of quantum entanglement:

- the entangled states are perfectly correlated, such that if the base used (or entangled pair) is, for example, that of Eq. (14.10) and both the sender and the receiver make local measurements on their respective particles, both can obtain

vertical (i.e., or $|0\rangle$) or horizontal (i.e., or $|1\rangle$) polarizations simultaneously with 100% probability, since when the base of Eq. (14.10) is measured it can only collapse to states $|00\rangle$ and $|11\rangle$ exclusively. However, the particular results are completely random, i.e., it is impossible for the sender (and for the receiver) to predict if the wave function will collapse to vertical polarizations or horizontal polarizations, and
- any measurement by a hacker destroys these correlations [44] in a way that the sender and the receiver can detect.

In the same way as in the case of the BB84 protocol, the E91 protocol involves a private measurement protocol before detecting the presence of the hacker. The measurement stage involves sender measuring each received photon using some bases from a set surely different from the receiver's chosen ones.

Both sender and receiver keep their respective series of basis choices private until measurements are completed. In this circumstances, two groups of photons are made:

- the first one consists of photons measured using the same bases chosen by the sender and the receiver, while
- the second one contains all other photons.

To detect the presence of a hacker, both sender and receiver can compute the test statistic using the correlation coefficients between sender's bases and receiver's similar to that shown in the Bell test experiments [47]. Using maximally entangled photons, such a statistic would result in $|S| = 2\sqrt{2}$. If not so, then both the sender and the receiver can conclude the hacker has introduced local realism to the system, violating Bell's Theorem [47]. If the protocol is correctly implemented, the first group can be used to generate keys since those photons are completely anti-aligned between the sender and the receiver.

14.5 Final Remarks

In this first section, the foundational concepts of Quantum Cryptography have been explored, as well as the two most relevant QKD exponents for public key distribution based on techniques own of Quantum Mechanics. The next section invites the reader to immerse themselves in the most modern techniques that will replace the previous ones thanks to their better performance and practicality, and which will be used in the most diverse applications proposed by the new millennium.

References

1. Zhou, T., et al, Quantum Cryptography for the Future Internet and the Security Analysis Hindawi Security and Communication Networks, Volume 2018, Article ID 8214619, https://doi.org/10.1155/2018/8214619
2. Buhrman, H., et al, Position-Based Quantum Cryptography: Impossibility and Constructions (2011) arXiv:1009.2490v4 [quant-ph]
3. Kollmitzer, C., and Pivk, M., eds, Applied Quantum Cryptography, Lecture notes in Physics 797, Springer, Berlin Heidelberg (2010)
4. Cariolaro, G.: Quantum Communications. Springer International Publishing, N.Y. (2015)
5. Mayers, D., Tourenne, C.: Violation of Locality and Self-Checking Source: A Brief Account. In: Tombesi P, Hirota O. (eds) Quantum Communication, Computing, and Measurement 3, 269–276. Springer, Boston, MA. (2002)
6. Yu, X.-T., Zhang, Z.-C., Xu, J.: Distributed wireless quantum communication networks with partially entangled pairs. Chin. Phys. B, 23:1, 010303. (2014)
7. NIST: Quantum Computing and Communication. CreateSpace Independent Publishing Platform. (2014)
8. Dieks, D.: Communication by EPR devices. Physics Letters A, 92:6, 271–272. (1982)
9. Phillips, A.C.: Introduction to Quantum Mechanics. Wiley, N.Y. (2003)
10. Bell, J.S.: Speakable and unspeakable in quantum mechanics. 52–62. Cambridge University Press, Cambridge. (2004)
11. ETSI: Quantum Key Distribution (QKD); Component characterization: characterizing optical components for QKD systems. ETSI GS QKD 011 V1.1.1. (2016)
12. Bechmann-Pasquinucci, H., Pasquinucci, A.: Quantum key distribution with trusted quantum relay. (2018) arXiv:quant-ph/0505089
13. Audretsch, J.: Entangled Systems: New Directions in Quantum Physics. Wiley-VCH Verlag GmbH & Co., Weinheim, Germany. (2007)
14. Jaeger, G.: Entanglement, Information, and the Interpretation of Quantum Mechanics. The Frontiers Collection. Springer-Verlag. Berlin, Germany. (2009)
15. Horodecki, R., et al., Quantum entanglement. (2007) arXiv:quant-ph/0702225
16. Abellán C, et al., Challenging local realism with human choices: The BIG Bell Test Collaboration. Nature, 557, 212–216. (2018)
17. Popescu S, Rohrlich D.: Causality and Nonlocality as Axioms for Quantum Mechanics. (1997) arXiv.org:quant-ph/9709026
18. Ghirardi, G.C., et al., Experiments of the EPR Type Involving CP-Violation Do not Allow Faster-than-Light Communication between Distant Observers. Europhys. Lett. 6, 2, 95–100. (1988)
19. Eberhard, P.H., Ross, R.R.: Quantum field theory cannot provide faster-than-light communication. Found. Physics Letters. 2, 2, 127–149. (1989)
20. Bancal, J.-D., et al., Quantum non-locality based on finite-speed causal influences leads to superluminal signaling. Nature Physics. 8, 867–870. (2012)
21. Herbert, N.: FLASH-A superluminal communicator based upon a new kind of quantum measurement. Found. Physics. 12, 12, 1171–1179. (1982)
22. Weinstein, S.: Superluminal Signaling and Relativity. Synthese, 148:2, 381–399. (2006)
23. Weinstein, G.: Einstein on the Impossibility of Superluminal Velocities. (2012) arXiv: physics.hist-ph/1203.4954
24. Kurtsiefer, C., et al., Entanglement enhanced quantum cryptography fiber systems, Long Distance Photonic Quantum Communication, QuComm Deliverable D17 (2002) https://pdfs.semanticscholar.org/e3ac/bfcd9893adba5c72e2c95e3e4c862842f26c.pdf
25. Ribordy, G., et al., Long distance entanglement based quantum key distribution, Phys. Rev. A 63, 012309 (2001)
26. Brassard, G., et al., Limitations on practical Quantum Cryptography, Phys. Rev. Lett. 85, 1330 (2000)

27. Mayers, D., Yao, A., Quantum Cryptography with imperfect Apparatus, Proc. of the 39th IEEE Conf. on Found. of Computer Science (1998)
28. de Riedmatten, H., et al., Report on entanglement enhanced quantum cryptography field trials, Long Distance Photonic Quantum Communica, QuComm Deliverable D22 (2003) http://citeseerx.ist.psu.edu/viewdoc/download?doi=10.1.1.559.3745&rep=rep1&type=pdf
29. Mastriani, M., Entanglement virtualization after the first quantum key teleportation <hal-02083356v3> (2019)
30. Mastriani, M., Is instantaneous quantum Internet possible? <hal-02161517v4> (2019)
31. Nielsen, M.A., Chuang, I.L., Quantum Computation and Quantum Information. Cambridge University Press, Cambridge. (2004)
32. Kaye, P., Laflamme, R., Mosca, M., An Introduction to Quantum Computing. Oxford University Press, Oxford. (2004)
33. Stolze, J., Suter, D., Quantum Computing: A Short Course from Theory to Experiment. WILEY-VCH Verlag GmbH & Co. KGaA. Weinheim, Germany. (2007)
34. Bennett, C.H., et al., Teleporting an Unknown Quantum State via Dual Classical and Einstein-Podolsky-Rosen Channels. Phys. Rev. Lett. 70, 1895. (1993)
35. Bouwmeester, B.D., et al., Experimental quantum teleportation, Phil. Trans. R. Soc. Lond. A, 356, 1733–1737. (1998)
36. Bouwmeester, D., et al., Experimental Quantum Teleportation. Nature, 390, 575–579. (1997)
37. Boschi, D., et al., Experimental Realization of Teleporting an Unknown Pure Quantum State via Dual Classical and Einstein-Podolsky-Rosen Channels. Phys. Rev. Lett., 80, 1121. (1998)
38. Kurucz, Z., Koniorczyk, Z., Janszky, J., Teleportation with partially entangled states. Fortschr. Phys. 49:10–11, 1019–1025. (2001)
39. Bennett, C.H.; Brassard, G. Quantum Cryptography: Public key distribution and coin tossing. Proc. of IEEE Intern. Conf. on Computers, Systems and Signal Processing, 175–179 (1984)
40. Tomamichel, M., Leverrier, A., A largely self-contained and complete security proof for quantum key distribution. Quantum. 1:14. (2017) https://doi.org/10.22331/q-2017-07-14-14
41. Portmann, C., Renner, R., Cryptographic security of quantum key distribution (2014) arXiv:quant-ph/1409.3525
42. Wootters, W.K., Zurek, W.H., A single quantum cannot be cloned. Nature, 299, 802–803. (1982)
43. Busch, P., Lahti, P., Pellonpää, J.P., Ylinen, K., Quantum Measurement. Springer, N.Y. (2016)
44. Schlosshauer, M., Decoherence, the measurement problem, and interpretations of quantum mechanics. Reviews of Modern Physics. 76:4, 1267–1305. (2005)
45. Ekert, A.K., Quantum cryptography based on Bell's theorem, Phys. Rev. Lett., 67:6, 661–663 (1991)
46. Ma, X-S., et al., Quantum teleportation using active feed-forward between two Canary Islands (2012) arXiv:1205.3909v1 [quant-ph]
47. Bell, J.S., On the Einstein-Podolsky-Rosen paradox, Physics, 1: 195–200 (1964)

Chapter 15
Quantum Tools

Contributed by Dr. Mario Mastriani
The content mentioned below in this chapter is contributed by Dr. Mario Mastriani based on the work conducted by him over the past decade. The authors of this book would like to thank Dr. Mastriani for his contribution and permissions to add this work in the book. As confirmed by Dr. Mastriani, the ideas specified below have not been published anywhere but are only discussed internally and stored in repositories like Academia.edu (https://www.academia.edu/39217563/Is_instantaneous_quantum_Internet_possible), and as the author and owner, he has the privilege to publish the content.

In this section will develop all the tools necessary to be able to implement the best Quantum Cryptography solutions involved in the most modern applications of the new millennium.

15.1   Quantum Entanglement

Let us suppose that we have two corrals with one cow in each, as shown in Figure 25.1. Each corral only has two gates: a spin-up and a spin-down. We open both gates in both corrals. The experiment begins with the corral on the left, where we have a 50% chance of the cow coming out of any of the gates. Besides, this experiment is consequent with the Bell's base of Eq. (14.10).

As we have mentioned in Subsection 24.4, the base of Eq. (14.10) can collapse to $|00\rangle$ or $|1\rangle$ when we make a local measurement on one of the two entangled particles. In such a way that if in the corral on the left we obtain a spin-up $|0\rangle$ as a result of the measurement we make, in the corral on the right we will obtain the same result, as shown in Fig. 15.1. But if in the corral on the left we get a spin-down $|1\rangle$, then in the corral on the right we will get the same result as well, i.e., $|1\rangle$, as we can see on Fig. 15.2. In both cases relative to the base of Eq. (14.10), nature, not

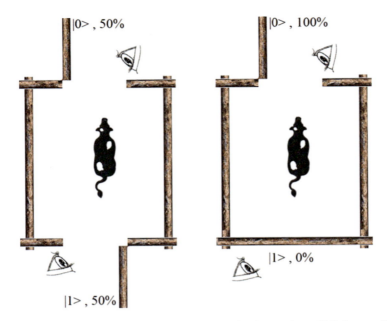

Fig. 15.1 The experiment begins with the corral on the left. That cow has a 50% chance of leaving through any of both gates: |0⟩ and |1⟩. When that cow decides to leave, e.g., through the spin-up gate |0⟩, then the cow from the corral on the right will exit through the spin-up gate |0⟩ too with a 100% of certainty. Nature, not us, takes control and blocks the spin-down gate |1⟩ in the corral on the right

us, takes control and blocks the opposite gate in the corral on the right. Besides, this happens instantaneously regardless of the distance between corrals.

That is, the decision of the cow on the left conditions the departure of the cow on the right, that is, we have no control over the outcome of the experiment. It is clear that we yield the control of the result of experiment to the cow on the left, instead of us. In other words, quantum entanglement is a random phenomenon which is instantaneously synchronized independently of the distance between entangled particles (Fig. 15.3).

The same experiment can be implemented on the Quirk platform [17] and is represented in Fig. 15.4. We have chosen this platform because it is the most pedagogical and has the best graphic expression of all simulators, however, this experiment can be reproduced with identical results on an optical circuit [18].

Figure 15.4 shows two qubits (q[0] and q[1]) and the generation of an Einstein-Podolsky-Rosen (EPR) pair thanks to the combination of a Hadamard's gate (H) and a CNOT gate. Quantum measurement (QM) in q[0] and q[1] represent a pair of detectors with vertical polarization. QM in q[0] represents the pair of eyes of the corral on the left of Fig. 15.1, while QM in q[1] represents the pair of eyes of the corral on the right of the same figure. DM means density matrix, Po|1⟩ means percent of |1⟩, and BS means Bloch's sphere. It is evident that the three witnesses

15.1 Quantum Entanglement

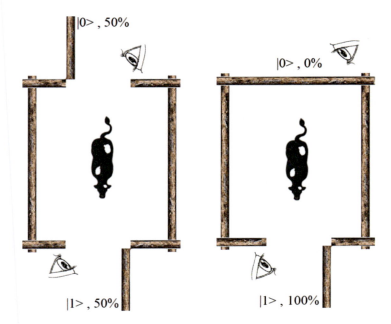

Fig. 15.2 The experiment begins with the corral on the left. That cow has a 50% chance of leaving through any of both gates: $|0\rangle$ and $|1\rangle$. When that cow decides to leave, e.g., through the spin-down gate $|1\rangle$, then the cow from the corral on the right will exit through the spin-down gate $|1\rangle$ too with a 100% of certainty. Nature, not us, takes control and blocks the spin-up gate $|0\rangle$ in the corral on the right

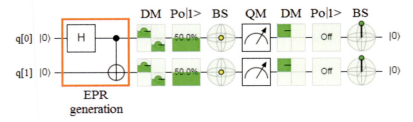

Fig. 15.3 Quirk implementation of the experiment of Fig. 15.1, where, DM, Po$|1\rangle$, BS, QM, mean Density Matrix, Probability of $|1\rangle$, Bloch's Sphere, and Quantum Measurement, respectively. QM in q[0] detects a spin-up, then instantaneously QM in q[1] detects a spin-up too, both with vertical polarization. Entanglement dies as a consequence of the measurement process

at the output of EPR generation block, i.e., DM, Po$|1\rangle$, and BS, show behavior in both qubits q[0] and q[1] typical of an EPR pair. Then, if QM in q[0] detects (continuing with the example of Fig. 15.1) a spin-up $|0\rangle$ for q[0], then q[1] resulting from QM in q[1] is absolutely defined as a spin-up $|0\rangle$ too, as this is evidenced by DM, Po$|1\rangle$, and BS at the output of both detectors. All the laboratory experiments carried out during the last decades support this result. Finally, Figure 25.4 represents

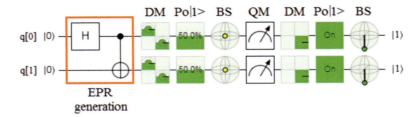

Fig. 15.4 Quirk implementation of the experiment of Fig. 15.2. QM in q[0] detects a spin-down, then instantaneously QM in q[1] detects a spin-down too, both with horizontal polarization. Entanglement dies again as a consequence of the measurement process

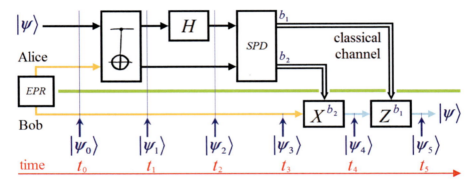

Fig. 15.5 Quantum teleportation using two classical bits for disambiguation

the implementation on Quirk platform [17] of the experiment of Figure 25.2, which is the counterpart to the previous one.

15.2 Quantum Teleportation

We will describe this protocol in a procedural way based on Fig. 15.5. First, an EPR pair is created and distributed between Alice and Bob. The horizontal green line of Fig. 15.5 separates Alice's and Bob's sides, and it can represent any arbitrary distance between them [5, 8–12, 19]. Besides, in Fig. 15.5 SPD block means single photon detector set, X, Z are Pauli's matrices activated by the bits b2, b1, respectively [2, 8–12], and EPR is the source of the entangled particles of Eq. (14.10). In fact, any of the EPRs of the Bell's basis complete set [2, 5] of the Eq. (14.10) can be chosen, however, we will use $|\beta_{00}\rangle = |\Phi^+\rangle$.

15.2 Quantum Teleportation

Alice's Side

1. Alice gets an arbitrary and unknown state to be teleported, $|\Psi\rangle = \alpha|0\rangle + \beta|1\rangle = \begin{bmatrix} \alpha & \beta \end{bmatrix}^T$ where $[\cdot]^T$ means transpose of $[\cdot]$ [31], and remembering from Subsection 24.2: $|\alpha|^2 + |\beta|^2 = 1$ and $\alpha \wedge \beta \in \mathbb{C}$ of a Hilbert's space [6].
2. $|\Psi\rangle = \alpha|0\rangle + \beta|1\rangle$ and $|\beta_{00}\rangle$ enter to a Bell State Measurement (BSM) block constitutes by a CNOT gate, a Hadamard's (H) gate and the SPD, then, a 3-partite state results in,

$$|\Psi_0\rangle = |\psi\rangle \otimes |\beta_{00}\rangle = |\psi\rangle|\beta_{00}\rangle = (\alpha|0\rangle + \beta|1\rangle)\frac{1}{\sqrt{2}}(|00\rangle + |11\rangle) =$$

$$\frac{1}{\sqrt{2}}[\alpha|0\rangle(|00\rangle + |11\rangle)] = \frac{1}{\sqrt{2}}[\alpha|000\rangle + \alpha|011\rangle + \beta|100\rangle + \beta|111\rangle] =$$

$$\begin{bmatrix} \frac{\alpha}{\sqrt{2}} & \frac{\beta}{\sqrt{2}} & 0 & 0 & 0 & 0 & \frac{\alpha}{\sqrt{2}} & \frac{\beta}{\sqrt{2}} \end{bmatrix}^T \quad (15.1)$$

where "$\otimes$" is the Kronecker's product [5]. For simplicity [6, 7], from now on, we will adopt $|x\rangle \otimes |y\rangle = |x\rangle|y\rangle$ in a generic form.

3. A CNOT gate is applied to $|\Psi_0\rangle$,

$$|\psi_1\rangle = \frac{1}{\sqrt{2}}[\alpha|000\rangle + \alpha|011\rangle + \beta|110\rangle + \beta|101\rangle] = \begin{bmatrix} \frac{\alpha}{\sqrt{2}} & 0 & 0 & \frac{\beta}{\sqrt{2}} & 0 & \frac{\beta}{\sqrt{2}} & \frac{\alpha}{\sqrt{2}} & 0 \end{bmatrix}^T \quad (15.2)$$

4. A Hadamard's (H) gate is applied to $|\psi_1\rangle$,

$$|\psi_2\rangle = \frac{1}{2}[|00\rangle X^0 Z^0 |\psi\rangle + |01\rangle X^1 Z^0 |\psi\rangle + |10\rangle X^0 Z^1 |\psi\rangle + |11\rangle X^1 Z^1 |\psi\rangle]$$

$$= \frac{1}{2}[|\Phi^+\rangle X^0 Z^0 |\psi\rangle + |\Phi^-\rangle X^1 Z^0 |\psi\rangle + |\Psi^+\rangle X^0 Z^1 |\psi\rangle + |\Psi^-\rangle X^1 Z^1 |\psi\rangle]$$

$$= \begin{bmatrix} \frac{\alpha}{2} & \frac{\alpha}{2} & \frac{\beta}{2} & \frac{-\beta}{2} & \frac{\beta}{2} & \frac{-\beta}{2} & \frac{\alpha}{2} & \frac{\alpha}{2} \end{bmatrix}^T$$

$$= \begin{bmatrix} \frac{\alpha}{2} & 0 & \frac{\beta}{2} & 0 & 0 & 0 & 0 & 0 \end{bmatrix}^T \rightarrow |\Phi^+\rangle \rightarrow |00\rangle \rightarrow 00 \rightarrow X^0 Z^0$$

$$+ \begin{bmatrix} 0 & \frac{\alpha}{2} & 0 & \frac{-\beta}{2} & 0 & 0 & 0 & 0 \end{bmatrix}^T \rightarrow |\Psi^+\rangle \rightarrow |10\rangle \rightarrow 10 \rightarrow X^0 Z^1$$

$$+ \begin{bmatrix} 0 & 0 & 0 & 0 & \frac{\beta}{2} & 0 & \frac{\alpha}{2} & 0 \end{bmatrix}^T \rightarrow |\Phi^-\rangle \rightarrow |01\rangle \rightarrow 01 \rightarrow X^1 Z^0$$

$$+ \begin{bmatrix} 0 & 0 & 0 & 0 & 0 & \frac{-\beta}{2} & 0 & \frac{\alpha}{2} \end{bmatrix}^T \rightarrow |\Psi^-\rangle \rightarrow |11\rangle \rightarrow 11 \rightarrow X^1 Z^1 \quad (15.3)$$

From the last four rows of the above equation, we can draw the following conclusions:

- they represent the quantum measurement carried out by Alice and the suggestion that she makes to Bob by transmitting the classical bits of disambiguation, by a classical channel, to reconstruct the teleported state [8–12] thanks to a pair of Pauli's matrices [5]: $X = \begin{bmatrix} 0 & 1 \\ 1 & 0 \end{bmatrix}$, and $Z = \begin{bmatrix} 0 & 1 \\ 0 & -1 \end{bmatrix}$
- the result of quantum measurement arises randomly and the four possible results are equally likely to appear
- the projections of the four bases are overlapping in relation to the detectors of the SPD [20], and
- given that the teleportation protocol begins with the distribution of an EPR pair, the quantum measurement kills the entanglement (i.e., it destroys the original arbitrary state on Alice's side) so as not to violate the No-Cloning Theorem [13], which implies that in order to teleport another state it must again distribute another EPR pair, and thus, the process continues indefinitely.

Figure 15.6 synthesizes the complete quantum teleportation protocol. With the same criterion as in the case of the EPR pair, we will analyze quantum measurement in the context of quantum teleportation through the example of the cow in the corral. We begin the experiment by leaving the four gates of the corral in Figure 25.6 open. The cow will have an identical probability of exiting by any of the four gates. Once the cow decides on a gate, e.g., $|00\rangle$, Alice observes it and transmits the result to Bob in the form of two classical bits ($b_1 = 0, and b_2 = 0$) sent by a classical channel, so that Bob can reconstruct the teleported state.

It is evident that the last four rows of Eq. (15.3), Figs. 15.6 and 15.7 are related to the point that the three, finally, represent the same thing, which consists that we exclusively have a state of observation without control over the experiment. As it will be seen in the next subsection, this is the reason for the need for a classical connection between Alice and Bob in addition to the EPR channel.

Meanwhile, we will implement the protocol of Fig. 15.5 on Quirk platform [17], in order to verify what has been said, although this protocol has been tested with

Alice's measurement	Alice transmits	This happens with probability	Collapsed state	Bob applies $X^{b_2}Z^{b_1}$
$\|\Phi^+\rangle \to 00$	$b_2 b_1 = 00$	$\left\|\frac{1}{2} X^0 Z^0 \|\psi\rangle\right\|^2 = \frac{1}{4}$	$\|\Phi^+\rangle X^0 Z^0 \|\psi\rangle$	$X^0 Z^0 \|\psi\rangle = \|\psi\rangle$
$\|\Psi^+\rangle \to 01$	$b_2 b_1 = 01$	$\left\|\frac{1}{2} X^0 Z^1 \|\psi\rangle\right\|^2 = \frac{1}{4}$	$\|\Psi^+\rangle X^0 Z^1 \|\psi\rangle$	$X^0 Z^1 \|\psi\rangle = Z\|\psi\rangle$
$\|\Phi^-\rangle \to 10$	$b_2 b_1 = 10$	$\left\|\frac{1}{2} X^1 Z^0 \|\psi\rangle\right\|^2 = \frac{1}{4}$	$\|\Phi^-\rangle X^1 Z^0 \|\psi\rangle$	$X^1 Z^0 \|\psi\rangle = X\|\psi\rangle$
$\|\Psi^-\rangle \to 11$	$b_2 b_1 = 11$	$\left\|\frac{1}{2} X^1 Z^1 \|\psi\rangle\right\|^2 = \frac{1}{4}$	$\|\Psi^-\rangle X^1 Z^1 \|\psi\rangle$	$X^1 Z^1 \|\psi\rangle = X Z\|\psi\rangle$

Fig. 15.6 Alice's side: measurement of the base, classical transmission of bits, and the collapse of states. Bob's side: classical reception of bits, gates application for the final recovery of the arbitrary state

15.2 Quantum Teleportation

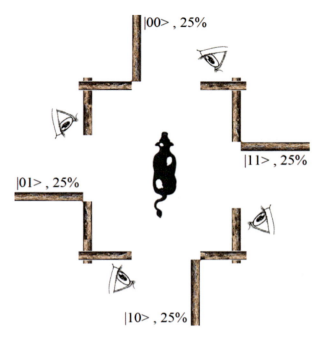

Fig. 15.7 The corral has the four gates open. The cow has the same chance of exiting through any of the four gates. When the cow decides to leave, e.g., through the $|00\rangle$ gate, then Alice observes and transmits the result to Bob in the form of two classical bits, in this case, $b1 = 0$, and $b2 = 0$

total success during the last two decades [9–11]. For both platforms, we choose the same state to be teleported, in order to compare outcomes, which results from:

$$q[0] = HTHS$$

$$q[0] = \frac{\begin{bmatrix} 1 & 1 \\ 1 & -1 \end{bmatrix}}{\sqrt{2}} \sqrt[4]{\begin{bmatrix} 1 & 0 \\ 0 & -1 \end{bmatrix}} \frac{\begin{bmatrix} 1 & 1 \\ 1 & -1 \end{bmatrix}}{\sqrt{2}} \sqrt[2]{\begin{bmatrix} 1 & 0 \\ 0 & -1 \end{bmatrix}} \begin{bmatrix} 1 \\ 0 \end{bmatrix} \quad (15.4)$$

$$= (85.3553\%|0\rangle, 14.6447\%|1\rangle)$$

with

$$H = \frac{\begin{bmatrix} 1 & 1 \\ 1 & -1 \end{bmatrix}}{\sqrt{2}}, T = \sqrt[4]{\begin{bmatrix} 1 & 0 \\ 0 & -1 \end{bmatrix}} = \sqrt[4]{Z}, S = \sqrt[2]{\begin{bmatrix} 1 & 0 \\ 0 & -1 \end{bmatrix}} = \sqrt[2]{Z}, q[0] = \begin{bmatrix} 1 \\ 0 \end{bmatrix}$$
(15.5)

Figure 15.8 shows the implementation of the quantum teleportation protocol of Fig. 15.5 on Quirk platform [17]. The blocks in color mean: orange for EPR generation, light blue for qubit generation (to be teleported), pink for BSM, gray for

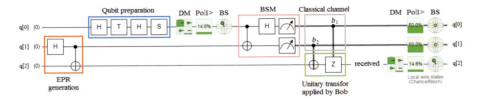

Fig. 15.8 Quantum teleportation implemented on Quirk [17] platform

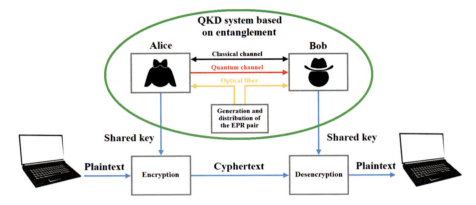

Fig. 15.9 Modern version of QKD based on entangled photon pairs

classical channel, and light green for the unitary transforms to be applied by Bob to recover the teleported state [8]. The complete coincidence between both sets of witness tools shows that the state is perfectly teleported.

15.3 Modern Version of QKD Based on Entangled Photon Pairs

In its most Laconic and Cartesian form, this version of QKD consists of five channels which can be grouped, in a functional way, in a minor number, i.e., there is reusability of the mentioned channels. Such channels, based on Fig. 15.9, are:

1. A link based on an optical fiber (bidirectional in yellow) to distribute the entangled photons which will constitute the inaccessible quantum channel (unidirectional in red)
2. Entangled photons constitute the inaccessible quantum channel (unidirectional in red) which vehicles the teleportation of the key, i.e., the public key with which the message is encrypted and decrypted
3. A classic disambiguation channel (gray block in Fig. 15.8, bidirectional in black in Fig. 15.9, which allows the reconstruction of the key on the receiver side

4. A classic channel (the same of the last item) to verify the integrity of the transmitted key, since as we have previously mentioned the key must be authenticated in QKD
5. A third, public, and classic channel (unidirectional in blue) for the transmission of the encrypted message or Cyphertext.

As we have mentioned before, the quantum channel is inaccessible for a hacker; however, the rest of the channel, i.e., the classics (including the optical fiber), are completely exposed to a hacker attack; for example, let us suppose that one of the classic bits transmitted from Alice to Bob via the classic channel (in gray) of Fig. 15.8 is maliciously altered, so according to Fig. 15.6, Bob will incorrectly reconstruct the message. This way both Bob and the hacker cannot get the correct message. For this reason we say that this configuration ensures the security of the message but not its integrity, which happens with all versions of QKD, no exceptions to date. In other words, while QKD uses classic channels for its operation, it will not constitute a perfect Quantum Cryptography system via public key distribution.

A very important detail to keep in mind is that both parties, sender and receiver, can share the key without problems, which implies that Alice keeps a copy of this key before teleporting it to Bob. Although this seems to violate the No-Cloning Theorem [13], it is not the case, since the keys are made up of CBSs, Eqs. (4) and (5), which can be cloned, compared [21], deleted [21], and survive intact quantum measurement [14].

15.4 Virtual Entanglement Procedure

In this section, a virtual entanglement procedure for a more practical QKD version is presented, where its most outstanding characteristics are:

1. It is a dynamic key cryptographic system, but unlike similar systems which work with a table of keys that are randomly selected at also random time intervals, in this case the current key is constructed from the previous key and the text plain, while encrypted text is built from the new key and plain text.
2. The procedure can be programmed so that the key changes with each character, word, or phrase, the most demanding being the case of change with each character and the least demanding the one of change with each phrase, with the exception that the more demanding the case, more computational resources will be consumed, therefore requiring a higher speed and quality network in order to avoid unnecessary latencies.
3. It is the first cryptographic system in history in which the first key can be openly reported to the hacker, in addition to being distributed between Alice and Bob, which synchronizes both ends.
4. The procedure is absolutely public, in fact it is published on the Internet [3, 4].
5. Each message begins with an incontextual character, word or phrase imposed by the same procedure, so that the hacker is working in brute force with such

an element while the entire message has already reached its destination in an indecipherable way.
6. The hacker has no possibility to get patterns and thus to do disambiguation between those patterns and then deduct the key. The hacker thus does not have enough information to proceed to the right cryptoanalysis technique.
7. It was created to be able to be implemented on non-secure networks of the type:
   - Electromagnetic link (e.g., radio)
   - Internet
   - Cellphone network
   - Networks of optical fiber
   - Satellite constellations.

In its most basic form the complete procedure consists of the following:

1. Alice generates the initial key k_0
2. Alice makes a copy of the key k_0
3. Alice teleports the key to Bob, then Alice and Bob share the initial key k_0
4. Alice applies an operator such that with the key k_0 and the Plaintext gets the following key k_1
5. Alice applies another operator such that with the key k_0 and the Plaintext gets the Ciphertext
6. Bob applies an operator such that with the key k_0 and the Cyphertext gets the following key k_1
7. Bob applies another operator such that with the key k_0 and the Cyphertext gets the Plaintext

Moreover, in every process of QKD, we identify two aspects, the tactical and the strategic:

- the tactical aspect is typically represented by the transmission of the key itself, which can be done by a quantum teleportation
- the strategic one, which puts the whole QKD procedure in a more general context.

This strategic aspect implies, in the most of the practical cases, the use of a classical channel for sending encrypted information, but with a series of important differences based on Fig. 15.10:

- we distribute an EPR pair, then the first and only key k0 is teleported safely and unalterably thanks to a quantum teleportation. The key can be at both ends of the link simultaneously without violating the No-Cloning theorem [13] because it is composed of CBSs exclusively, which unlike generic qubits can be cloned, compared [21], and deleted [21], in fact, CBSs remain unchanged after a quantum measurement [14].
- On Alice's side:
  - with the plaintext PT_i and the previous key k_{i-1} function F builds the new key k_i

15.4 Virtual Entanglement Procedure

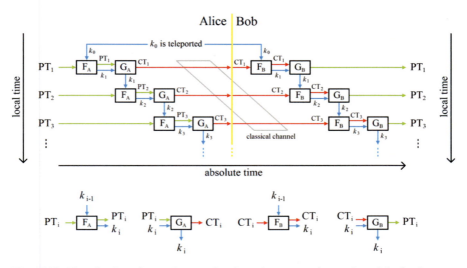

Fig. 15.10 Virtualization of entanglement after the real quantum teleportation of the first key

- with the plaintext PT_i and the new key k_i function G builds the ciphertext CT_i,
- On Bob's side:
  - with the ciphertext CT_i and the previous key k_{i-1} function F builds the new key k_i
  - with the ciphertext CT_i and the new key k_i function G rebuilds the plaintext PT_i.

It is evident that there is an entanglement between functions F and G on both sides of the classical channel according to the four equivalent bases of Bell in this new context:

$$\begin{aligned}
|B_{FF}\rangle &= |F^A \cap F^B\rangle \cup |G^A \cap G^B\rangle \\
|B_{FG}\rangle &= |F^A \cap F^B\rangle \sqcup |G^A \cap G^B\rangle \\
|B_{GF}\rangle &= |F^A \cap G^B\rangle \cup |G^A \cap F^B\rangle \\
|B_{GG}\rangle &= |F^A \cap F^B\rangle \sqcup |G^A \cap G^B\rangle
\end{aligned} \quad (15.6)$$

where, $a \cap b$ means interaction between a and b, $c \cup d$ means complementation between c and d, and, $c \sqcup d$ means anti-complementation between c and d, i.e., they are not the known operators of union and intersection in sets theory. The action of the operators, interaction, complementation, and anti-complementation, allows a satisfactory encryption and a perfect recovery of the plaintext in a perfect deterministic synchrony, which raises the level of sensitivity of the channel, in such a way that an apocryphal presence would be detected as a lack of coherence in the received message.

That elevation in the sensitivity level imitates the action of an observer in a quantum measurement context, where we must remember that every quantum measurement alters what is measured. In other words, we are creating an artificial entanglement and we are keeping it alive by the action of the operators $\{\cap, \cup, \underline{\cup}\}$ and functions $\{F, G\}$ as if the flow of QKD was never interrupted.

When we say that the functions $\{F, G\}$ work with the plaintext or the ciphertext, in reality what we mean is that F and G are able to use: everything, part or a characteristic of said texts within each function. The most innocent of such a feature could be some amount of something (e.g., sum of any character of the texts). Anyway, we must understand that by putting all this in its corresponding context, i.e., quantum Internet, the information will definitely be organized in frames and layers (and quantum Internet protocol datagrams), so that we will have a series of other tools to play with.

When a hacker acts in the middle of the classical channel, which does not have the functions F, G or the strategy between them, i.e., the correct used base $\{|B_{FF}\rangle, |B_{FG}\rangle, |B_{GF}\rangle, |B_{GG}\rangle\}$, and tries to measure the message, the result of that measurement will be like a random martingale, which also emulates the random result of a quantum measurement on an entangled particle. In addition, decoherence in the plaintext recovered by Bob will indicate an apocryphal presence in the channel. Although the analysis of Bob can be done on a witness element, which is complementary to the plaintext and before obtaining it, which can act as a flag indicating the integrity of the procedure.

Since the whole procedure depends, in some way, on a whimsical creation of the message to be transmitted, the practical effect of the whole procedure is equivalent to the permanent change of the encryption and decryption algorithms with each message. Moreover, we can configure the procedure so that everything changes with each frame or subframe. Therefore, users should not memorize keys nor think new ones, being the security transparent to all of them, because nobody knows its key and nevertheless it always changes capriciously and randomly with each frame (or another think controlled by clocks to both sides of classical channel) of the message, i.e., permanently.

For all that said, and although the procedure in Fig. 15.10 is about the teleportation of keys, it is in some way also a mechanism for teleportation of messages with important information.

Perhaps the least intuitive of the three operators $\{\cap, \cup, \underline{\cup}\}$ is that of anti-complementation, freedom, or compensation, however, in all three operators, we leave each future user free to select the finest identity and combination of them to obtain the best result according to the use they want to give them.

What about the impact of noise on the procedure in Fig. 15.10? This question has two answers:

- classic: the same impact that any classical communication system has, for which we can resort to the whole arsenal of tools tending to mitigate errors such as error correcting codes, use of parity bits, etc.,
- quantum: the same impact expected in a quantum system when we must coexist with noisy gates and decoherence as a result of quantum measurement [14, 15] at the end of the process.

This configuration is recommended for use in the event that the transport network is of questionable security, considering that it must be used by the financial and banking system for Quantum Blockchain [22–24], Quantum Money [25–27], and Quantum Check [28–30]. Finally, there is no doubt its impact on Quantum Internet [31–36], and therefore on Quantum IoT.

15.5 Quantum Internet

In order to have a considerably more secure Internet for all types of data exchange, quantum Internet [31–36] was born, which consists of a complete set of tools such as:

- entanglement swapping for quantum repeaters,
- quantum repeaters for quantum Internet, and
- quantum memories for buffering.

Next, the three tools will be described.

15.5.1 Entanglement Swapping for Quantum Repeaters

Entanglement swapping [37–43] is the technique traditionally used in quantum repeaters [1, 44, 45], evidently from quantum teleportation [8–12]. Here, we will present both versions:

- Entanglement swapping, properly speaking
- Teleportation Swapping, i.e., the exchange of two teleported qubits between Alice and Bob in order to try bidirectionality within the quantum teleportation.

Entanglement Swapping
Figure 15.11 represents the entanglement swapping thanks to an implementation of the technique on Quirk [17], where, on the left side of that figure, we can see the generation of two pairs of entangled particles. The first EPR pair results from q[0]

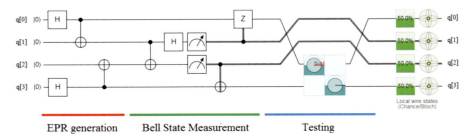

Fig. 15.11 Entanglement swapping on Quirk platform [17]

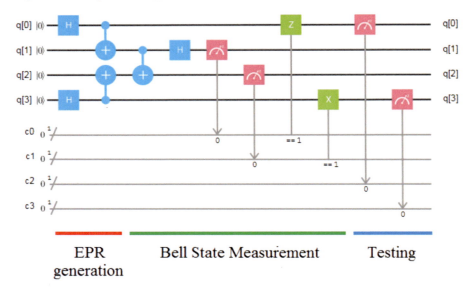

Fig. 15.12 Entanglement swapping on IBM Q platform [46]

and q[1] qubits, which are first at ground state |0⟩. The second EPR pair arises from q[2] and q[3] qubits, also at ground state at the beginning.

After the blocks that generate the EPR pairs, we can see a Bell State Measurement (BSM) block. Finally, to the right of Fig. 15.11, qubits q[0] and q[3] end up being entangled. Instead, Fig. 15.12 shows the same configuration of Fig. 15.5 but implemented on IBM Q [46]. Figure 15.13 represents the distribution of probabilities for this experiment where the results (o outcomes) were: $0.132 + 0.126 + 0.124 + 0.109 = 0.491$ 0.5 for |00⟩ on the right, and $0.133 + 0.131 + 0.124 + 0.121 = 0.509$ 0.5 for |11⟩ on the right side too, with the following experiment parameters: shots = 1024, and seed = 7. Although the action of decoherence is evident, given that the outcomes are not exactly 0.5, however, the results obtained are quite good. We must bear in mind that in every real physical implementation, a very important factor in relation to quantum Internet will appear being the coherence

15.5 Quantum Internet

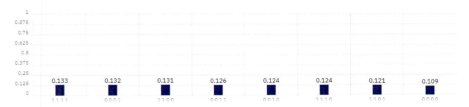

Fig. 15.13 Distribution of probabilities for the experiment of Fig. 15.12

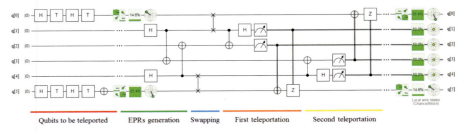

Fig. 15.14 Teleportation swapping on Quirk platform [17]

time of the EPR pairs, and their relation with the distances to be covered during the distribution of these, which will further degrade the outcomes.

Teleportation Swapping

As we have mentioned before, this technique represents an effort to give bidirectionality to quantum teleportation (Fig. 15.14). That is Alice and Bob interexchange qubits at the end of the process. Figure 25.13 represents the implementation of this protocol on Quirk platform [17]. Now, let us suppose that Alice generates a qubit q[0] = (85.4% |0⟩, 14.6% |1⟩), and Bob generates another qubit q[5] = (14.6% |0⟩, 85.4% |1⟩). Then we have a block for the generation of two EPR pairs and then another module with a pair of SWAP gates. The following two quantum teleportations will cross-transmit the corresponding qubits in such a way that at the end of the whole process Alice will receive in q[0] the qubit generated by Bob, while Bob will receive in q[5] the qubit generated by Alice. Figure 15.15 shows the protocol implemented on IBM Q platform [46], while Fig. 15.16 represents the probability distribution for the experiment of Fig. 15.15.

15.5.2 Quantum Repeaters for Quantum Internet

Although quantum Internet is currently automatically associated with the distribution of EPR pairs over terrestrial fiber optic networks, in this work we have opted for the satellite version, since we understand that it is the future of Quantum Internet. Therefore, in this subsection we will present three protocols of quantum repeaters, namely:

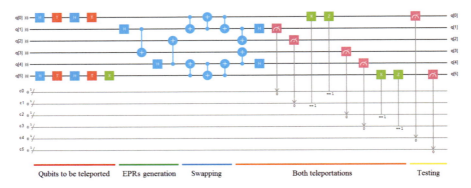

Fig. 15.15 Teleportation swapping on IBM Q platform [46]

Fig. 15.16 Probability distribution for the experiment in Fig. 15.15

- Relay race or satellite posts: it is a sequential protocol, which allows us to chain satellites in order to take the qubit to be teleported to any arbitrary distance
- Teleportation of EPRs: in this protocol we first teleport an EPR pair (which arise from a pivot point) to two satellites in two diametrically opposed locations of the space, so that at one end we introduce the qubit to be teleported, while at the other end we receive it. It is a parallel system, which works across the width in order to divide the distribution time of the EPRs by half, taking better advantage of the coherence time. On the other hand, multiplying the number of pivot points, the range has no limits, assuring us that the complete procedure will always be within the coherence time. Finally, the pivot points can be on a planet, or on other satellites
- Hybridization: It is a mix between the previous two versions

Relay Race or Satellite Posts

This protocol is clearly sequential and works just like a relay race, and just like in rugby the pass of information is always forward, in an exclusive manner. According to Fig. 15.17, which represents the implementation of this protocol on Quirk platform [17], the qubit q[0] to be teleported is prepared to act on point (1). The first qubit teleporter puts the qubit on point (2), then the second qubit teleporter puts the qubit on point (3), i.e., the final destination on q[4].

Notwithstanding the above, the segmentation of the quantum circuit that we can see in the lower part of Fig. 15.17 is not entirely accurate. The reality is that we must perform a regrouping of blocks more in line with reality according to the real responsibility of each actor involved: Alice on land, quantum satellite, and Bob on

15.5 Quantum Internet

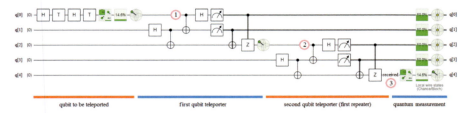

Fig. 15.17 Satellite post on Quirk platform [17]

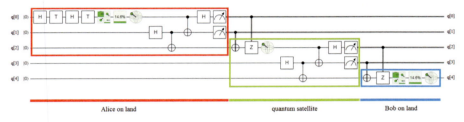

Fig. 15.18 Reconfiguration of the layout belongs to the satellite post on Quirk platform [17]

land, where the latter is obviously in a different location and it is very distant from Alice. We will only perform this regrouping for this case in order not to overload or complicate the subsequent figures corresponding to other protocols. This regrouping can be seen in detail in Fig. 15.18. The first macro-block highlighted in red includes: the qubit to be teleported, the first EPR generation and distribution, and the first BSM which will transmit the disambiguation bits to the quantum satellite via a radio channel. This first macro-block will be labeled as Alice on land. The second macro-block highlighted in light green and labeled as quantum satellite includes: a unitary transformation thanks to which we will reconstruct the teleported state from an element of the first EPR pair and the classical bits received by radio, and the generation of a second entangled pair and a second BSM, which will transmit the disambiguation bits to Bob on land via another radio channel. Finally, the third macro-block highlighted in blue and labeled as Bob on land includes a second unitary transformation, which received the second element of the second EPR pair as well as the disambiguation bits thanks to which Bob can reconstruct the teleported state.

Figure 15.19 allows us to see a practical representation of the regrouping of Fig. 15.18 with a clear correspondence between all the involved blocks. All begin on point A (on land), where the block labeled as BSM is inside macro-block highlighted in red in Fig. 15.18, while the second BSM block will be on board inside quantum satellite. The point B (on land) receives an element of the EPR pair and the classical bits of disambiguation emitted by the quantum satellite with which Bob applies a unitary transformation and thus reconstructs the teleported state.

Figure 15.20 represents the same protocol of Fig. 15.17 but on IBM platform, while Fig. 15.20 shows the probability distribution for this experiment with the

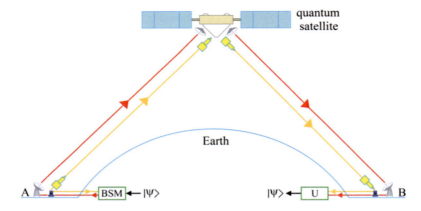

Fig. 15.19 Detailed allocation of all elements of Fig. 15.18

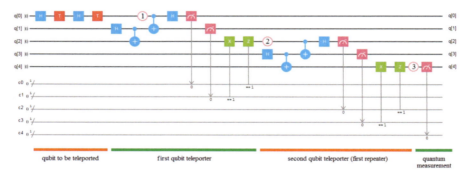

Fig. 15.20 Satellite post on IBM Q platform [46]

Fig. 15.21 Probability distribution for the experiment of Fig. 15.20

next result: q[4] = (84.78% |0⟩, 15.22% |1⟩) where the experiment parameters were: shots = 1024, seed = 7 (Fig. 15.21).

Teleportation of ERPs

This protocol begins with the distribution of an EPR pair from a point called pivot (P). This pivot (P) is on the orange segment corresponding to the generation of the EPR pair, which can be seen on the left side of the Quirk implementation of Fig. 15.22. Each of the elements of the EPR pair will be teleported by one of the two satellites.

15.5 Quantum Internet

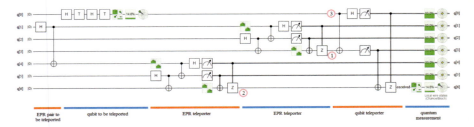

Fig. 15.22 Teleportation of EPRs on Quirk platform [17]

Said satellites will be located at an equidistant distance from the pivot (P) at opposite ends and will be responsible for teleporting the EPR pair to their two ground destinations, points (1) and (2), both on land, which make up the two ends of the communication link. Subsequently, the element of the EPR pair in point (1) and the qubit to be teleported in point (3) enter the BSM on the right of Figure 25.21 on the segment labeled as qubit teleporter, whereupon the teleportation process begins. At the other end of the link (also on the ground) two sets of unitary transformations reconstruct the teleported qubit. The main reason for this protocol is to halve the distribution time of the EPRs to try to make the most of the coherence time. It is a parallel and simultaneous procedure, however, Fig. 15.22 does not do justice to the true protagonists of this protocol. Instead, Fig. 15.22 shows that in this protocol there are actually five very well-defined actors, which are macro-blocks labeled as:

- Alice on land (in red): this macro-block includes the qubit generation block (i.e., the qubit to be teleported), and the last BSM for the final qubit teleportation
- EPRs to be teleported (in green): it is simply a block for the generation of an EPR pair, which are distributed from land
- Quantum satellites (in blue): they are two, each with its own block of EPR pair generation, and BSM, and
- Bob on land (in yellow): which includes two sets of unitary transformations for the reconstruction of the teleported state and the final measurement process.

The reconfiguration of the layout which belongs to the teleportation of EPRs on Quirk of Fig. 15.22 can be understood better from Fig. 15.23, where we can see: the pivot (P) on the north pole of Earth, both quantum satellites, and the points A (emitter) and B (receiver). The last unitary transformation at the point B receives the classical bits for disambiguation from Internet. Both quantum satellites contain their own EPR pair generation and BSM blocks (Figs. 15.23, 15.24, and 15.25).

The teleportation of EPRs on IBM Q platform can be seen in Fig. 15.20. It is an identical scheme to that of Fig. 15.22 for the Quirk platform. Figure 15.26 shows the probability distribution for this experiment with the next result: q[4] = $(83.19\%|0\rangle, 16.80\%|1\rangle)$, where the experiment parameters were: shots = 1024, seed = 7. The obtained results are clearly of inferior quality to those of Figs. 15.20 and 15.19 for the case of satellite post, which is reasonable

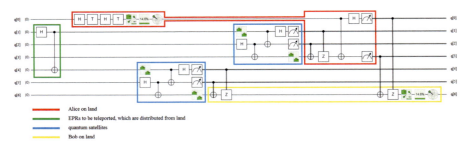

Fig. 15.23 Reconfiguration of the layout belongs to the teleportation of EPRs on Quirk [17]

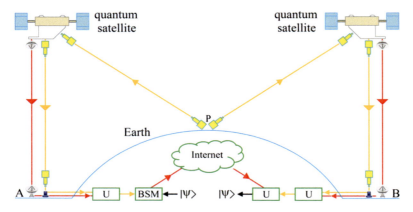

Fig. 15.24 Detailed allocation of all elements of Fig. 15.23

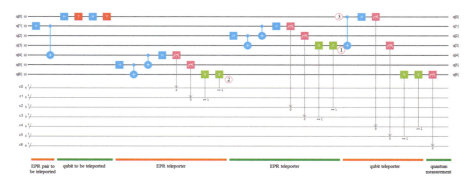

Fig. 15.25 Teleportation of EPRs on IBM Q platform [46]

Fig. 15.26 Probability distribution for the experiment of Fig. 15.25

because in that experiment five quantum measurements were used, while in this experiment with teleportation of EPRs, as shown in Fig. 15.25, we had to use seven. The increase in the quantum measurement number automatically increases decoherence [14, 15].

Hybridization

As we mentioned at the beginning of Sect. 15.5, hybridization consists of merging the two techniques discussed above: satellite post and teleportation of EPRs. Figure 15.27 represents the hybridization on Quirk platform. The main difference between this layout and that one present in Fig. 15.22 is the duplication in the number of satellites that teleport each element of the EPR pair which arises from the pivot point (P). In other words, each EPR is sequentially teleported by two quantum satellites instead of one, thus multiplying by 2 the range of said distribution. Comparing the qubit q[0] to be teleported on the left-up corner of Fig. 15.27 with q[10] on the right-down corner of the same figure, we can see that the teleportation was completely successful.

The hybridization on IBM Q platform [46] is presented in Fig. 15.28. We can notice the absence of a figure with the probability distribution. Due to the huge number of bars of that figure, we will only give the final outcome: q[10] = 87.4% |0⟩ 12.6% |1⟩). The presence of 11 quantum measurements explains the deterioration of the teleported qubit compared to the original value. The experiment parameters were: shots = 1024, seed = 7.

Finally, in the case of Fig. 15.28 was impossible to obtain the probability outcomes as in the previous cases due to the excessive number of qubits, in fact, 10, which eliminated any possibility that the IBM Q platform would give us the results.

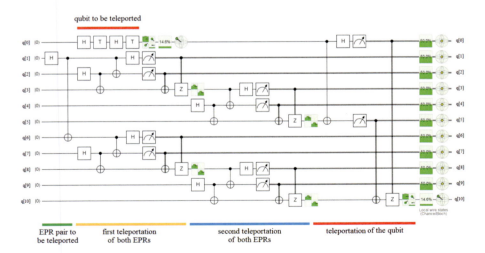

Fig. 15.27 Hybridization on Quirk platform [17]

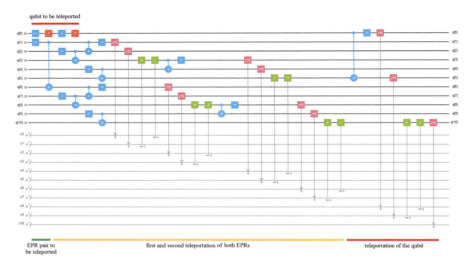

Fig. 15.28 Hybridization on IBM Q platform [46]

15.5.3 Quantum Memories for Buffering

They represent the quantum version of the classical ones which store quantum states instead classical bits. In the quantum Internet context, they are employed to make temporary buffering of an entangled photon while its counterpart travels to the receiver. Its main application is to generate the proper delay so that both entangled photons are available at the exact moment. They are present in each BSM. Therefore, their use in practice is necessary both for quantum teleportation [8–12, 16, 19, 20] and quantum Internet [31–36].

15.6 Final Remarks

In this section, we have explored the most modern tools associated with Quantum Cryptography, which will be of fundamental importance when dealing with the applications that the new century brings us, in particular, all those associated with the data security of the most critical systems, i.e., financial, military, and government.

References

1. Bechmann-Pasquinucci, H., Pasquinucci, A.: Quantum key distribution with trusted quantum relay. (2018) arXiv:quant-ph/0505089
2. Audretsch, J.: Entangled Systems: New Directions in Quantum Physics. Wiley-VCH Verlag GmbH & Co., Weinheim, Germany. (2007)
3. Mastriani, M., Entanglement virtualization after the first quantum key teleportation <hal-02083356v3> (2019)
4. Mastriani, M., Is instantaneous quantum Internet possible? <hal-02161517v4> (2019)
5. Nielsen, M.A., Chuang, I.L., Quantum Computation and Quantum Information. Cambridge University Press, Cambridge. (2004)
6. Kaye, P., Laflamme, R., Mosca, M., An Introduction to Quantum Computing. Oxford University Press, Oxford. (2004)
7. Stolze, J., Suter, D., Quantum Computing: A Short Course from Theory to Experiment. WILEY-VCH Verlag GmbH & Co. KGaA. Weinheim, Germany. (2007)
8. Bennett, C.H., et al., Teleporting an Unknown Quantum State via Dual Classical and Einstein-Podolsky-Rosen Channels. Phys. Rev. Lett. 70, 1895. (1993)
9. Bouwmeester, B.D., et al., Experimental quantum teleportation, Phil. Trans. R. Soc. Lond. A, 356, 1733–1737. (1998)
10. Bouwmeester, D., et al., Experimental Quantum Teleportation. Nature, 390, 575–579. (1997)
11. Boschi, D., et al., Experimental Realization of Teleporting an Unknown Pure Quantum State via Dual Classical and Einstein-Podolsky-Rosen Channels. Phys. Rev. Lett., 80, 1121. (1998)
12. Kurucz, Z., Koniorczyk, Z., Janszky, J., Teleportation with partially entangled states. Fortschr. Phys. 49:10–11, 1019–1025. (2001)
13. Wootters, W.K., Zurek, W.H., A single quantum cannot be cloned. Nature, 299, 802–803. (1982)
14. Busch, P., Lahti, P., Pellonpää, J.P., Ylinen, K., Quantum Measurement. Springer, N.Y. (2016)
15. Schlosshauer, M., Decoherence, the measurement problem, and interpretations of quantum mechanics. Reviews of Modern Physics. 76:4, 1267–1305. (2005)
16. Ma, X-S., et al., Quantum teleportation using active feed-forward between two Canary Islands (2012) arXiv:1205.3909v1 [quant-ph]
17. https://algassert.com/quirk
18. A. Furusawa, and P. van Loock, Quantum Teleportation and Entanglement: A Hybrid Approach to Optical Quantum Information Processing, Wyley-VCH, Weinheim, Germany (2011)
19. Zeilinger, A., Quantum teleportation and the non-locality of information, Phil. Trans. R. Soc. Lond. A, 355, 2401–2404. (1997)
20. Herbst, T., et al., Quantum teleportation over a 143 km free-space link, International Conference on Space Optics (ICSO'2014) Tenerife, Canary Islands, Spain (2014)
21. Arul, A.J., Impossibility of comparing and sorting quantum states. (2001) arXiv:quant-ph/0107085
22. Rajan, D., Visser, M., Quantum Blockchain Using Entanglement in Time (2019) arXiv:1804.05979v2 [quant-ph]
23. Hasnain, Md., et al., Demonstration of quantum blockchain, and therein CNOT and ping-pong attacks on IBM QX (2020) https://www.researchgate.net/publication/338571490
24. Banerjee, S., Mukherjee, A., Panigrahi, P.K., Quantum Blockchain using Weighted Hypergraph States (2019) https://www.researchgate.net/publication/337199562
25. Radian, R., Sattath, O., Semi-Quantum Money (2019) arXiv:1908.08889v4 [quant-ph]
26. Brodutch, A., et al., An adaptive attack on Wiesner's quantum money (2014) arXiv:1404.1507v4 [quant-ph]
27. Moulick, S.R., Panigrah, P.K., Signing Perfect Currency Bonds (2015) arXiv:1505.05251v2 [quant-ph]
28. Diep, D.N., van Minh, N., Quantum E-Cheques (2017) arXiv:1705.10083v1 [quant-ph]

29. Behera, B.K., Banerjee, A., Panigrahi, P.K., Experimental realization of quantum cheque using a five-qubit quantum computer, Quantum Inf. Process., 16,312 (2017)
30. M. Caleffi, A.S. Cacciapuoti, F.S. Cataliotti, S. Gherardini, F. Tafuri and G. Bianchi, The Quantum Internet: Networking Challenges in Distributed Quantum Computing. arXiv:quant-ph/1810.08421v2 (2019)
31. M. Caleffi, A.S. Cacciapuoti and G. Bianchi, Quantum Internet: from Communication to Distributed Computing. arXiv:quant-ph/1805.04360v1 (2018)
32. L. Gyongyosi and S. Imre, Entanglement Access Control for the Quantum Internet. arXiv:quant-ph/1905.00256v1 (2019)
33. L. Gyongyosi and S. Imre, Opportunistic Entanglement Distribution for the Quantum Internet. arXiv:quant-ph/1905.00258v1 (2019)
34. T. Satoh, S. Nagayama, T. Oka and R. Van Meter, The network impact of hijacking a quantum repeater, IOP Quantum Science and Technology, 3:3, 034008 (2018) https://doi.org/10.1088/2058-9565/aac11f
35. A.S. Cacciapuoti, M. Caleffi, R. Van Meter, L. Hanzo, When Entanglement meets Classical Communications: Quantum Teleportation for the Quantum Internet (Invited Paper). CoRR abs/1907.06197 arxiv:quant-ph/1907.06197 (2019)
36. Zukowski, M., et al., Event-ready-detectors bell experiment via entanglement swapping. Phys. Rev. Lett. 71, 4287–4290. (1993)
37. Pan, J.-W., et al., Experimental entanglement swapping: Entangling photons that never interacted. Phys. Rev. Lett. 80, 3891–3894. (1998)
38. Jennewein, T., et al., Experimental nonlocality proof of quantum teleportation and entanglement swapping. Phys. Rev. Lett. 88, 017903. (2001)
39. Tsujimoto, Y., et al., High-fidelity entanglement swapping and generation of three-qubit GHZ state using asynchronous telecom photon pair sources. Scientific Reports, 8:1446. (2018)
40. Jin, R.-B., et al., Highly efficient entanglement swapping and teleportation at telecom wavelength. Scientific Reports 5, 9333. (2015)
41. Schmid, C., et al., Quantum teleportation and entanglement swapping with linear optics logic gates. New Journal of Physics 11, 033008. (2009)
42. de Riedmatten, H., et al., Long-distance entanglement swapping with photons from separated sources. Phys. Rev. A 71, 050302. (2005)
43. Lloyd, S., et al., Long Distance, Unconditional Teleportation of Atomic States via Complete Bell-state Measurements, Phys. Rev. Lett. 87, 167903. (2001)
44. Sangouard, N., et al., Quantum repeaters based on atomic ensembles and linear optics. Rev. Mod. Phys., 83:33-80. (2011)
45. https://quantum-computing.ibm.com/
46. Zhang, R., Xue, R., Liu, L., Security and Privacy on Blockchain (2019) arXiv:1903.07602v2 [cs.CR]

Chapter 16
Applications

Contributed by Dr. Mario Mastriani
The content mentioned below in this chapter is contributed by Dr. Mario Mastriani based on the work conducted by him over the past decade. The authors of this book would like to thank Dr. Mastriani for his contribution and permissions to add this work in the book. As confirmed by Dr. Mastriani, the ideas specified below have not been published anywhere but are only discussed internally and stored in repositories like Academia.edu (https://www.academia.edu/39217563/Is_instantaneous_quantum_Internet_possible), and as the author and owner, he has the privilege to publish the content.

There have been many applications that have been identified that use Quantum computing. In this chapter, we will discuss a selected list of the most conspicuous and representative applications of the Quantum Cryptography techniques analyzed so far, namely:

- Quantum Internet of Things (QIoT)
- Quantum Blockchain
- Quantum Money and Quantum Cheque
- Quantum Security of Confidential Documents
- Quantum Radar
- Quantum Tomography.

16.1 Quantum Internet of Things (QIoT)

It is the set of techniques for controlling all types of remote devices using the techniques seen in the previous sections, in particular quantum Internet based on quantum teleportation and QKD based on entangled photon pair. In the latter case,

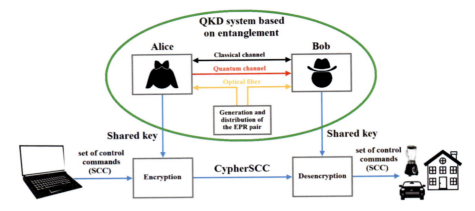

Fig. 16.1 A basic scheme of QKD based on entangled photon pair applied to IoT. In this scheme, the Alice and Bob labels exist for the sole purpose of giving identity to both ends of this configuration, taking as reference the original functionality of QKD

in the scheme of Fig. 16.1, similar to that of Figure 25.8, the plaintext consists of a set of control commands for different devices at home and office.

However, the same technique has an evident application in smart cities for the intelligent use of all the city's common resources: supply of electricity, gas, water, and Internet; traffic lights; monitoring of nuclear power plants; among many others. The security system aims to prevent terrorist attacks, making control and monitoring commands inaccessible over both a shared or restricted network. For this reason, there is an evident need for an optical fiber cable or a satellite system such as the one mentioned above to properly serve this application, which can be exploited by a smart city hub.

16.2 Quantum Blockchain

Formally, a blockchain is a growing list of blocks that are linked using cryptography. Each block contains a cryptographic hash of the previous block, a timestamp, and transaction data (generally represented as a Merkle tree) [46]. For this reason, Blockchain adapts perfectly in its implementation to two of the tools seen above:

- virtual entanglement procedure
- QKD based on entangled photon pair

The green oval of Fig. 16.2 shows the teleportation of the first hash of the quantum Blockchain, i.e., k_0, since this is the only hash that needs to be teleported, so from here we move from the architecture of Fig. 16.2 to the automatic procedure from where with the hash of the current block k_{i-1}, and the content of the current block $b_i \equiv PT_i$, we generate the hash of the next block k_i, and with the content of

16.3 Quantum Money and Quantum Cheque

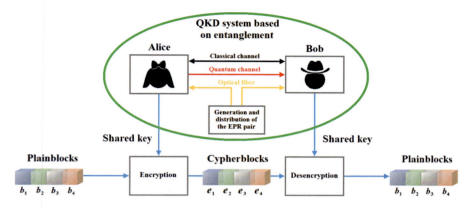

Fig. 16.2 QKD based on entangled photon pair giving rise to an improved version of Blockchain, from a much more efficient process of generating the hash of each block associated with the previous block, based on the architecture of which by means of the hash of the current block k_{i-1}, and the content of the current block $b_i \equiv PT_i$, we generate the hash of the next block k_i, and with the content of the current block and the hash of the next block k_i we encrypt the current block $b_i \equiv PT_i$ getting $e_i \equiv CT_i$

the current block $b_i \equiv PT_i$ and the hash of the next block k_i we encrypt the current block $b_i \equiv PT_i$ getting $e_i \equiv CT_i$.

This simple but extremely efficient procedure frees us from the annoying situation of depending on the quality of the hash, since here the hash of the next block is built with the hash of the current block and the whimsical and random content of the current block. This automatically reinforces the structural weaknesses of the original Blockchain by dramatically raising the level of security for the entire architecture.

16.3 Quantum Money and Quantum Cheque

This scheme is much simpler than the previous case since it simply consists of the provision of a secure link to protect transactions in which data related to credit cards and checks are involved.

Figure 16.3 shows this architecture which allows us to make all type of transactions via the typical classical channel, that is, Internet, and among which we can mention the purchase of goods, the payment of services, as well as any other type of commercial or financial operation, which includes deposits and transfers.

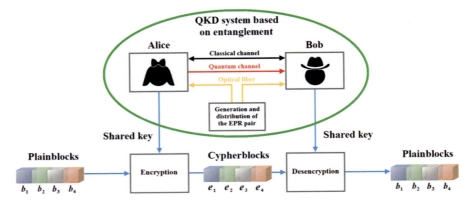

Fig. 16.3 QKD based on a pair of entangled photons applied to the purchase of goods, the payment of services, as well as any other type of commercial or financial operation, which includes deposits and transfers via Internet

16.4 Quantum Security of Confidential Documents

This configuration, which can be seen in Fig. 16.4, allows us to transmit all kind of graphics material: plans, certificates, photos, reserved material, graphics evidence involved in a legal case, among many others. This scheme is absolutely compatible with all international recognized standards of digital signature.

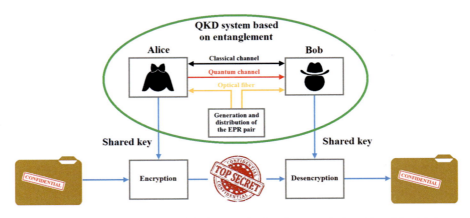

Fig. 16.4 QKD based on an entangled photon pairs for the transmission of confidential documents: plans, certificates, photos, reserved material, graphics evidence involved in a legal case, and so on

16.5 Quantum Radar

Based on the quantum teleportation protocol, see Fig. 16.5, quantum radar converts half of the entangled photons into signals of very high frequencies, of the order of X/Gamma rays fired at the target, with operational architecture like that of Fig. 16.6.

The entangled photon of the upper branch impacts to the target $|\Psi\rangle$. The rebound in the target is measured in the same branch, where the ambiguity described arises. This ambiguity is eliminated thanks to the measurement made in the Bell's State Measurement (BSM) block. The result of this measurement is expressed as two classic bits that appear randomly, which are transmitted to the lower branch (via a classic channel) in order to activate a couple of Pauli's matrices in the unit transformation labeled as U. Finally, after applying this transformation, we will obtain a visual version of the target.

While a common radar bases its target detection capacity on two fundamental aspects: (a) that the wavelength of the emitted signal is of the order of wavelength of the target's geometry, and (b) that an echo is essential, which in the case a stealth

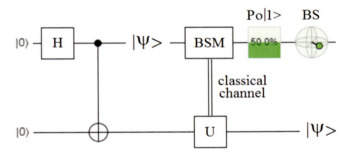

Fig. 16.5 Quantum radar interpreted as a quantum teleportation protocol, where the combination of the gates H (Hadamard) and CNOT creates and distributes an EPR pairs

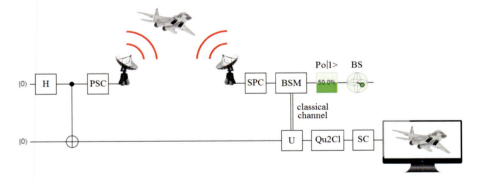

Fig. 16.6 Visual Detection and Ranging (VIDAR) based on a framework typical of a quantum teleportation

plane is noticeably dimmed or removed; quantum radar emits a signal of such a short wavelength to the target that it practically bounces off the fuselage particles at the molecular level generating the much desired echo, albeit immersed in a sea of undesirable noise due to atmospheric aspects. As a direct consequence of its operation, this type of radar generates an image very similar to a noisy ultrasound instead of a point on a screen as in the case of a common radar, for this reason it acts as a Visual Detection and Ranging (VIDAR), which after the appropriate treatment for image improvement allows to perform an Automatic Target Recognition (ATR) of the target for its subsequent interception.

The combination of the H (Hadamard) and CNOT gates creates and distributes EPR pairs. The entangled photon of the upper branch is converted to a very high frequency signal, in the order of the X/Gamma rays, thanks to a Photon to Signal Converter (PSC). The wavelength of the signal is so small that it impacts at an almost molecular level with the material of the stealth plane fuselage and recovers an echo that could not be obtained with normal radar. In this way the main advantage of the stealth plane is eliminated. The echo is converted from signal to photon thanks to a Signal to Photon Converter (SPC), to be measured in the BSM. With a procedure similar to that of Fig. 16.5, the qubits recovered from the unitary transform of the lower branch pass through a Quantum-to-Classical interface, and after that by a Signal Conformer (SC) to end up being shown on a monitor. The image of said monitor is only schematic, since the true image that is seen is more similar to a kind of ultrasound with a high noise content. However, once said noise has been processed, VIDAR can detect morphology and thus activate an Automatic Target Recognition (ATR).

The need for security on the part of VIDAR with architectures shown above, is evident given that the echo detected is generally transmitted to a command and control center in a different location than VIDAR, and by the way, very distant.

16.6 Quantum Medical Imagery

With an operating scheme similar to the previous case, although with much less noise, the only notable difference with quantum radar from the operational point of view lies in the transducer. Quantum tomography uses a transversal quantum illumination, with which we will obtain the target $|\Psi\rangle$ in real time, known as slice, which will allow us to make *ipso facto* diagnoses in the same place of the instrument or at great distances from its location. One of the most common practices in which this technology will be involved is known as robotic telesurgery.

The need for data security is the same as in the case of quantum radar since it is very common for the hospitals that carry out the study to send the images to:

- a local medical imagery storage system
- a national and/or international repository of tomographies
- the same patient

16.7 Final Remarks

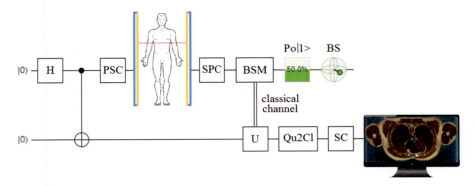

Fig. 16.7 Quantum tomography with a scheme of working identical to the previous case but with a difference, the transducer in the upper branch, which allows the signal to cross through the patient's body generating a laminar looking. Finally, the monitor displays the corresponding slice for real-time diagnosis at the same location or remotely

- other hospitals for research in a framework of cooperation, consultation, or second opinion (Fig. 16.7).

16.7 Final Remarks

As we could see, the scheme discussed under quantum cryptography can be adopted to all the data security needs that we have mentioned. In this context, the architecture is the ideal complement to the QKD based on an entangled photon pairs, as was clearly evidenced in the quantum Blockchain as discussed in this chapter. Finally, both techniques constitute the pair of new weapons in the Quantum Cryptography arsenal of the new millennium.

Index

A
Access control lists (ACLs), 20
Additive homomorphic encryption, 80–81
Advanced Encryption Standard (AES), 55, 57
AES 256, 55
Apple iPhone, 13
Application layer, 30
Artificial intelligence (AI), 9, 47, 69
Asymmetric encryption, 93
Asymmetric key cryptography
  advantages of, 64
  drawbacks of PKI, 64
  public key cryptosystem, 63–64
Audio technique, 41–42
Augmented reality (AR), 12
Authentication requirements of devices, 33
Automatic Target Recognition (ATR), 148
Availability, 48

B
Ballot, 73–74
Bell's bases, 112
Bell State Measurement (BSM) block, 123, 125, 132, 135, 137, 140, 147, 148
BGV scheme, 86
Big data, 10, 34, 69
Bloch's sphere, 109, 110
Block ciphers
  AES, 57
  cipher block chaining, 57
  cipher feedback mode, 57
  counter mode, 57
  DES, 57
  Galois/Counter Mode, 57
  Initialization Vector, 57, 58
Bluetooth, 21, 28
Bluetooth Low Energy (BLE), 28, 39–40
Bootstrapping, 18, 19, 21
  definition, 37
  existing bootstrapping techniques, 42
  existing out-of-band communication methods, 38
  OOB
    audio technique, 41–42
    Bluetooth Low Energy, 39–40
    haptics/touch, 40
    magnetic field technique, 41
    vibration technique, 42
    visual field technique, 41
  trusted device, 38
Botnets, 17, 22–23
Brute force attacks, 59
Buffering, quantum memories for, 140
Business layer, 32

C
C/C++ based library, 85–88
"Ceasar Cipher," 48
Certificate revocation lists (CRLs), 64
Certificates, 21, 37, 51, 64, 146
Certification Authority (CA) certificates, 37
Chosen ciphertext attacks (CCA), 53
Cipher block chaining (CBC), 57
Cipher feedback mode (CFB), 57
Ciphertext, 52–53, 79
Cloud computing, 72, 76

CoAP, 28
Computational Basis States (CBS), 110, 111
Confidentiality, 48
Counter mode (CTR), 57
Cryptanalysis, 48, 59–60
    differential cryptanalysis, 60
    differential power analysis, 60
    known plaintext attack, 60
Crypto algorithm, 81
Cryptographic keys, 39, 60
Cryptography
    current state-of-the-art, 49
    history, 48
Cryptosystems, 49, 51, 53
Crypto tokens, 58
Cyber-physical systems (CPSs), 3, 9, 10, 27

D
Data analysis, 72–73
Data Encryption Standard (DES), 55, 57
Data security in cloud, 70–72
Decryption, 48, 79
Denial of Service (DoS), 18, 41
Device lifecycle management, 34
Differential cryptanalysis, 60
Differential power analysis, 60
Diffie-Hellman, 64
Digital signatures, 33, 51, 64, 146
Distributed Denial of Service (DDoS), 18
Distributed phase reference coding, 109

E
Eavesdropping, 40, 64
ECC, 64
Einstein-Podolsky-Rosen (EPR) pair, 120
    teleportation of, 136–139
Electoral systems, 73–74
ElGamal encryption scheme, 75, 81–82
Empty pulses, 108
Encryption, 48
Entangled photon pairs, QKD, 114–115, 126–127
Entanglement swapping, 131–133
ENVEIL, 87

F
FHEW, 85
Fitness tracking, 14
Fully homomorphic encryption (FHE), 76

G
Galois/Counter Mode (GCM), 57
Goldwasser-Micali method (GM method)
    analysis of, 79
    decryption, 79
    encryption, 78–79
    generation, 78
    randomness, 77
Google glasses, 12
GPRS, 11

H
Hadamard's gate, 120, 123
Hajime botnet, 22, 23
Haptics/touch, 40
Hilbert's space, 110, 111
Homomorphic encryption, 52–53
    advantages, 83
    ballot and electoral systems, 73–74
    data security in the cloud, 70–72
    ElGamal encryption scheme, 81–82
    fully homomorphic encryption, 76
    GM method, 77–79
    industrial involvement
        challenges, 85
        ENVEIL, 87
        HElib, 86
        libraries, 85
        PALISADE, 87
        SEAL, 86
    Paillier scheme, 80–81
    partially homomorphic encryption, 75
    problems of importance in PKI, 76
    RSA cryptosystem, 82–83
    somewhat homomorphic encryption, 75
    support to data analysis, 72–73
    timeline, 78
Homomorphic encryption code, 100
Homomorphic Encryption library (HElib), 86
Horizontal polarizations, 122
HVAC system, 20
Hybridization, 139–140

I
IBM Q platform, 140
IBM's HElib, 86
Implementation
    code details, 99–101
    evaluation, 103
    performance, comparison of, 101–103
    Tamper evident, 103

Index

Initialization Vector (IV), 57, 58
Integrity, 48
Interface or API protection, 35
Internet-of-things (IoT), 47
   applications, 27
   bootstrapping (*see* Bootstrapping)
   business layer, 32
   communication mode, 28
   definition, 27
   five-layer version, 30, 31
   large-scale adoption, 29
   narrow waist of the current internet protocol, 29
   overall working, 28
   possible attacks, 17, 18
   processing layer, 31
   security
      access control, 20–21
      authentication, 19–20
      botnets, 22–23
      challenges to securing smart devices, 21
      device bootstrapping, 18–19
      techniques to protect devices, 23
   six layers in IoT security, 32
      authentication requirements of devices, 33
      device lifecycle management, 34
      encrypting data to prevent leakage of information in clear text, 33–34
      interface or API protection, 35
      securing the network of operation, 33
      storage solutions, 34
   smart-light system, 30
   three-layer version
      application layer, 30
      network layer, 30
      perception layer, 30
   transport layer, 31
iPhone, 13
IP subnet, 30

K
Known plaintext attack, 60
Kronecker's product, 111

L
LIDARs, 39
LoRA, 21, 28, 31
LTE modules, 11, 31

M
Machine learning, 47, 69, 104
Magnetic field technique, 41
Man-in-the-middle (MITM) attack, 42
Mckincey Global Institute, 11
Microsoft SEAL, 86, 87
Mirai, 22, 23
Multi-factor authentication (MFA), 58

N
Nanorobots, 7
Network layer, 30
NFC, 31
No-Cloning theorem, 124, 128
Non-malleable encryption, 91, 93, 99, 100

O
One-Time Password (OTP), 33
Out-of-band (OOB) channels
   audio technique, 41–42
   Bluetooth Low Energy, 39–40
   haptics/touch, 40
   magnetic field technique, 41
   vibration technique, 42
   visual field technique, 41

P
Paillier encryption scheme
   analysis of, 80
   benefits of additive homomorphic encryption, 80–81
   PHE scheme, 80
PALISADE, 87
Partially homomorphic encryption (PHE), 75, 80
Passive attacks, 64
Perception layer, 30
Personal Identification Number (PIN), 33
Photon pairs, 108
Photon polarization state, 113
Photon to Signal Converter (PSC), 148
Pivot, 136
Polarized single photon, 112–114
Private key, 63, 64
Processing layer, 31
Public key cryptosystem, 63–64
Public key encryption techniques, 92–93
Public Key Infrastructure (PKI), 37, 49–51, 64
Python wrappers, 87–88

Q

Quality of Experience (QoE), 14, 51
Quality of Service (QoS), 14, 51
Quantum Bit Error Rate (QBER), 109
Quantum blockchain, 144
Quantum cheque, 145
Quantum cryptography, 107, 143
    QKD
        entangled photon pairs, 114–115
        polarized single photon, 112–114
    quantum information processing, primer on, 111–112
    quantum key distribution, prolegomenous on, 108–109
Quantum entanglement, 108, 119–122
Quantum indeterminacy, 113
Quantum information processing, primer on, 109–112
Quantum internet
    buffering, quantum memories for, 140
    entanglement swapping, for quantum repeaters, 131–133
    quantum repeaters for, 133–140
Quantum Internet of Things (QIoT), 131, 143
Quantum key distribution (QKD)
    entangled photon pairs, 114–115, 126–127
    on polarized single photon, 112–114
    prolegomenous on, 108–109
Quantum measurement (QM), 120
Quantum mechanics, 108
Quantum medical imagery, 148
Quantum memories, 140
Quantum money, 145
Quantum radar, 147–148
Quantum repeaters, 131–133
Quantum satellite, 134, 137
Quantum security of confidential documents, 146
Quantum teleportation, 122–126
Quantum tomography, 148
Quantum tools
    QKD, entangled photon pairs, 126–127
    quantum entanglement, 119–122
    quantum internet
        buffering, quantum memories for, 140
        entanglement swapping, for quantum repeaters, 131–133
        quantum repeaters for, 133–140
    quantum teleportation, 122–126
    virtual entanglement procedure, 127–131
Qubit basis, 110
Qubits, 120
Quirk platform, 120, 125, 135, 137, 139

R

RADARs, 39
Radio frequency identification (RFID), 11
Ransomware attack, 22
Relay race, 134
RFID, 31
RSA cryptosystem, 50, 64, 82–83

S

Satellite posts, 134–136
Sensor-infused smart-home, 7
Sensors
    automation of processes, 7
    classification technique, 4
    need for, 6
    in smartphone, 5
    wireless sensor networks, 5–6
Sensory nodes, 27, 28, 30, 31
Service Set Identifier (SSID), 33
Shodan, 20
Signal Conformer (SC), 148
Signal to Photon Converter (SPC), 148
Simple Encrypted Arithmetic Library (SEAL), 86, 87
6LoWPAN, 28
Smart card, 33
Smart devices
    augmented reality, 12
    categorization, 9
    CPSs, 9, 10
    data generation, 15
    devices for monitoring health, 13
    google glasses, 12
    handheld device, 11–13
    IoT paradigm, 11
    sensors, 13–14
    virtual reality, 12
    wearable computing, 11–14
Smart-homes sensors, 7, 8
SMS, 33
Somewhat homomorphic encryption (SHE), 75
SONARs, 39
State-of-art public key encryption techniques, 92
Stream ciphers, 55
SWAP gates, 133
Symmetric encryption, 93
Symmetric key cryptography
    advantages, 55
    applications, 55
    block ciphers
        AES, 57
        cipher block chaining, 57

Index

cipher feedback mode, 57
counter mode, 57
DES, 57
Galois/Counter Mode, 57
Initialization Vector, 57, 58
cryptanalysis, 59–60
    differential cryptanalysis, 60
    differential power analysis, 60
    known plaintext attack, 60
disadvantages, 58
key search (brute force) attacks, 59
stream ciphers, 55
systems based attack, 60
Systems based attack, 60

T
Tamper evident, 103
Tamper evident solutions (TED-SP)
    advantages, 94–95
    information storage and retrieval applications, 95
    proposed switchable-malleable encryption approach, 93, 94
    working of the proposed scheme, 94
TCP/IP model, 29
TEDSP, 99, 101, 102
Telemetry, 14
Teleportation, 136–139
Teleportation swapping, 133
Telnet ports, 23
Transport layer, 31
Transport Layer Security (TLS) protocol, 55
Trust bootstrapping, 42
Trusted device bootstrapping, 38

U
Ultrasonic sensors, 39

Unique identifiers (UIDs), 6
Unsecure IoT devices, 23
U.S. National Institute of Standards and Technology (NIST), 57
US National Science Foundation (NSF), 10

V
Vertical polarizations, 113–115
Vibration technique, 42
Virtual entanglement procedure, 127–131
Virtual reality (VR), 12
Visual Detection and Ranging (VIDAR), 147, 148
Visual field technique, 41
VLAN, 30
VPN, 23

W
Wearable computing, 3, 11
Web-of-Trust (WoT) model, 37
WEP, 34
Wi-Fi, 39
Wireless Fidelity (Wi-Fi), 28
Wireless sensor and actuation networks (WSANs), 6
Wireless sensor networks (WSNs), 3, 5–6
Worldwide Threat Assessment, 13
World Wide Web (WWW), 11
WPA, 34
WPA2, 34

Z
Zigbee, 28
Zombies or bots, 22
ZWave, 21, 28

Printed by Books on Demand, Germany